```
          S
          Y
          N
WATERCOURSE
          N
          Y
          M
          S
```

COLIN F. BROWN

Even queried suggestions have been given, in the belief
that mere flashes of thought by an expert may often
point the way towards correct findings.

From Calum Macpharlain's editorial note in the
1911 printing of Alexander Macbain's
Etymological Dictionary of the Gaelic Language.

To all past and present scholars, associated
with my particular field of research,
this book is gratefully dedicated.

CONTENTS

Preface

This book, which has over 2000 entries, is a companion to the Eilert Ekwall *English River-Names*, published in 1928, still considered as the standard work of reference for river-names despite its age. The collection presented here, hopefully, will set a standard in much the same way as the Ekwall treatise because, as far as I am aware, nothing comparable to it has been published to date. Overall, it is the culmination of 20 years' research into watercourse synonyms, the generic terms which define water flowing in a natural or artificial channel – which have appeared in print – in one form or another, in the national languages of the United Kingdom of Great Britain and Northern Ireland, the Irish Republic and the Isle of Man, over the last two millennia. Namely: English and Scottish; the P- Celtic group – Cornish, the cognate Breton taken to France and the Welsh; the Q- Celtic group – Irish, Manx and Scottish Gaelic. Other languages spoken here include Anglicized Irish; Anglo-Norman; Anglo-Saxon (given as Old-English); Gypsy; Latin and the Norn of Orkney and Shetland plus the problematical Pictish. In addition, a good number of cognate forms, found in Danish, Dutch, Faroese, Flemish, French, German, Icelandic, Norwegian, Swedish and even Old Prussian, are also given, due to the fact that they all share a common language history with, in particular, Old and Modern English, as we all belong to the Indo-European group of languages, which is where a great number of our root words come from.

This book does not attempt to present every form of every synonym found, or claim to be a complete survey of *all* the generic terms, but, it does include a good number of historically interesting forms, where possible, because, more often than not, they have turned out to be relative to current research.

There are some really amazing spellings and derivations to be found in the works of past authorities; place-name books, in particular, always seem to throw up some of the most bizarre spellings, derivations and the like. Watercourse generics, over time, have amassed thousands of different forms – river has toted up about 100 – and we have such diversities as: a brone for burn; a rune for rhine and a puylle for pill! The avon family has produced at least 120 variants including river-name forms, and, there are records to be found in place-names too; local historian Lester Steynor, collected a record 161 spellings for Bromsberrow, in Gloucestershire, during the 90's. River-name enthusiasts are well catered for too and need look no further than the River Wharfe, in Yorkshire. A really fantastic spelling for this name is given in the *Coucher Book of The Cistercian Abbey Of Kirkstall,* p 139 as: "Kuuerfe", in fact, the Wharfe river-name variants have some of the most unbelievable spellings – no less than seven different initial letters have been used for the name – H, K, O, Q, T, U and W, spanning the last 1000 years, from various printed sources. Derivations have often been the subject of much speculation and debate; sometimes based on wild guesses or extravagant theories, such as the one advanced by William Hutchinson (2:450), writing on *The History of the County of Cumberland,* in 1794: "We are induced to believe, the etymology of names having *Thor* or *Thur* in them, being taken for holy places consecrated to the Saxon deity *Thor,* are very erroneous. *Thur* is a Danish word, and signifies a brook or rivulet …. The contrary would take away the Danish derivation which we have adopted." Arguments for and against a particular definition can last for years; one instance, that immediately springs to mind is the contentious river-name Avon, initially given as a generic term, by the old masters, as a synonym for 'water, river'; rejected subsequently by a number of 20th century scholars, only to end up now widely accepted by the majority!

The entries concentrate on the vast amount of material available, from the earliest Anglo-Saxon charters and Latin documents; dictionaries; glossaries; gazetteers; maps; place-name books; classical sources and the like, right through to the ever increasing, highly popular, internet e-resources, constantly updated and now set firmly in place, in so many fields, as the ultimate tool of research. The leading authorities and dedicated experts in the fields of cartography, etymology, geography, hydronymy, history, lexicography, philology, topography and toponymy have, over time, presented us with a wealth of material and a platform for the contemporary researcher to work on, without which, this book would be very much the poorer. To this we can add, criticism, often levied on even the most expert of authorities, is not welcome here, as the author sincerely believes that, although very often superseded by contemporary scholarship, the fruits and labours of past professionals, in the works they presented, were, at the time, given in good faith, the most important thing of all being – what *is* there – rather than what is not!

All synonyms presented herein have been checked for mistakes, but, I would add to this that any omissions, imperfections or errors in the work, are mine alone. Naturally, with any book of this kind, there is always room for improvement – no doubt much has been missed – parts of the composition overlooked; queries left unanswered – hence the epigraph – consequently, any additions or amendments to the entries would, of course, be most welcome, thus enabling a greater enhanced future edition.

No matter what your age or status, there is something here for everyone and a wealth of information can, no doubt, be gleaned from the bibliographic entries.

I can only hope that the foregoing is tempting enough for the reader to carry on and discover a few gems and surprises along the way in my *magnum opus*.

Colin F Brown
Gornal Wood
England
2015

Acknowledgements

As far as I am aware, all references are fully credited to the respective authors without plagiarism, breach of copyright and the like; quotations from works currently *in* copyright have been permitted following requests by the author to the respective authorities.

I am greatly indebted, in particular, to Professor Gregory Toner, Project Director of the electronic Dictionary of the Irish Language [eDIL]; Christine Robinson, MA PhD, Director of the Scottish Language Dictionaries, DSL, [DOST & SND] and, Andrew Hawke, Managing Editor of the University of Wales, Dictionary of the Welsh Language, *Geiriadur Prifysgol Cymru* [GPC], who, in their wisdom, have kindly and readily given me permission to quote from their monumental works of reference, without which, and, without doubt, would render this collection of synonyms very much the poorer. The Irish eDIL e-resource; the collective DSL volumes which cover the *Dictionary of the Older Scottish Tongue*, DOST, and the *Scottish National Dictionary*, SND, together with the electronic version of the GPC, now form *the* standard works of reference for Old Irish, Scottish and Welsh words online – a truly magical resource for researchers worldwide.

I am also greatly indebted to The Council of The Early English Text Society [EETS] who have graciously given me permission to quote words (mainly Latin and Middle English) from their publications, which have certainly enhanced my historical word form entries for the Latin and Middle English periods in particular, and, at the same time, alleviated a shortfall in presentation.

Special thanks also, to Dr Ken George, for allowing me to quote from his Cornish dictionaries, *Gerlyver Kernewek Kemmyn - An Gerlyver Kres* and *An Gerlyver Meur;* works, which have certainly enriched and elevated my own Cornish listings.

In addition, I am extremely grateful to the various editors and authors of the leading online e-resources and journals, namely – Professor Gerhard Köbler, of the University of Innsbruck, for permission to quote word forms from his Germanic dictionaries: the Old English; Old Low Frankish; Old Frisian; Old High German; Gothic; Old Norse; Old Saxon plus the Germanic and Indo-Germanic dictionaries are the finest contemporary collection available; the use of which has allowed me to present a greatly enhanced root list. The Bosworth-Toller *Anglo-Saxon Dictionary and Supplement,* hosted by the Faculty of Arts, Charles University, Prague, is *the* best user friendly version currently available, permitting widespread use of many Old English terms, some of which would otherwise have to be omitted; Dr L Palmaitis and P Holcwesscher, for open access and citations from the *Old Prussian Elbing Vocabulary* and Chris Bond for permission to quote from his Cornish word lists. The United States Board on Geographic Names (USBG) deserve special commendation for making their online resources readily available; one of the greatest collections of geographic information available anywhere on the WWW.

According to the Academic Rights & Journals Department at the Oxford University Press, the use of the Oxford English Dictionary [OED] is very restricted; OED1 is not in copyright and is, therefore, free to use. OED2 is the 20 volume, 2nd edition, published in 1989 and OED3, is the 3rd edition, now online; available by subscription, or public library card log in. The 2nd and 3rd editions are copyrighted works and as such any quotations or WWW presentations are strictly prohibited without hefty fees. However, for the purpose of this work, most of the entries in OED1 have remained unchanged since 1888, and although later editions have amendments and additions in some cases, they do not significantly change the definitions or word forms of watercourse synonyms listed in the original edition.

The Middle English Dictionary [MED], hosted by the University of Michigan, is a copyrighted work, which requires permission before use. Words in the synonym list, headwords in particular, have been used subject to the *Fair Use Evaluation Document,* under Section 107 of the U.S. Copyright Code, provided by the University, following my permission request. Part of the permission agreement is to protect third party use, so, under no circumstances should any of the MED be used, in a commercial project, without first obtaining permission from the copyright office.

Finally, I shall always be eternally grateful to my dear wife Lorraine, who has encouraged me from the start and proof read the entire work at various stages; a very time-consuming activity, which definitely accelerated completion of the manuscript.

It is understandable that not all publishers choose to permit free access to their works, as the effort, cost and time-consuming activities needed to produce works such as the OED can take several years. My compilation of *Watercourse Synonyms* has been very time-consuming too, so, in view of this, and the statements above regarding permissions, I would, therefore, ask that prospective authors, planning a commercial project for publication, seek permission and credit by name and page any extracts they wish to use. It's a simple request, but one very often overlooked as I have discovered during my own work on the bibliographic side of things – very easy to spot as I have often come across certain extracts that have been used word-for-word without due credit being given! Although we are allowed to use and cite OED1 entries, it is much more rewarding to obtain the original sources used therein, hence the size of the bibliography, because, more often than not, the 'extras' gained override the effort needed to obtain them. I use the OED1 after exhaustive searches for originals have failed, usually for 17th century earlier works, variant forms or, in some cases, etymological sequences, unavailable from reference libraries or internet resources.

Notes on the Entries

All entries are arranged alphabetically, irrespective of bracketed or compounded forms, which may be hyphenated or spaced; dashes; diacritics; diphthongs and ligatures etc., such as the æ ligature, originally representing a Latin diphthong. In the Norse languages; particularly, the Icelandic and Faroese, *á* follows *a*, in their alphabet. Special characters in Danish, Norwegian and Swedish follow z. For the purpose of this work, the order will be *a* followed by *á*, followed by the Danish, Norwegian and Swedish *å*, followed by the English *aa* and finally the Norwegian *åa*. The OE special characters, ð or đ and þ, follow t in the alphabet.

Headwords are presented in lower case bold type, followed by a bracketed abbreviated language list; etymologies and roots, where possible; definitions; variant forms; bibliographic coded sources; cross references; relative place-name sources and suggestions for further reading, where appropriate, as illustrated in the example below.

beck (E, G, S Nn. & Sc.) From the ON bekkr, 'a brook or stream', very common in the eastern and northern counties of England settled by the Danes and Norwegians ... Parish (17) in his *Sussex Dialect* book gives "a rivulet". The OE forms are "bec, becc and bece", B-T (67 & 74); becc is also found in the ME *Cursor* Mundi – Morris (1874:2:515 line 8946). PP (29b) renders beck – "Bek watyr, rendylle, *riuulus*." WW (736.24) lists: "Hec rivulus, a bek."... See bache; backie; all the beck- stems; Macbain (1922:82) for the PN's Diebek and Dubec; The EDD (213a) for the dialect regions and, for lists of associated PN's, EPNE (1:26).

Bibliographic references, within the entries, use the 'author or work – date – volume – page – column - system', with line numbers added in some cases. For one author, one work for example: Parish (17). For authors or works with multi-volumes: Morris (1874:2:685a, line 11942). Some works, but very few, published in the same year, necessitate the addition of an a, b or c added to the date, for a single work example: Lindsay (1921b:179 line 320). Any multi-volume work would, of course, include a volume number too.

Double quote marks ("…") are used to distinguish a source quote from an author quote or definition which uses single marks ('…').

In the notes which follow, square brackets are used for the bibliographic references.

Notes on the Languages

There are still many questions to be answered regarding the arrival of the various languages that have reached these shores over the last 3000 years. However, for the purpose of this book, only the words quoted have any currency. Scrutinizing the works of Jackson [LHEB], [Lockwood], [Price] and [Oppenheimer], just to name a few, will soon convince the reader just how big and widespread the problem really is – the chronological events spanning language arrivals are well and truly in dispute, previous notions having been blown away in recent years, due partly to genetic analysis, and are, therefore, outside the scope of this book, other than for a brief summary by some of our early writers. One of the most interesting accounts is given by Androw of [Wyntoun] in his *The Orygynale Cronykil of Scotland*, c1400, in which he states: (lines 1373-6) "Off Langagis in Bretayne sere I fynd, that sum tym fyff thare were: Off Brettys fyrst, and Inglis syne, Peycht, and Scot, and syne Latyne." In other words – British; English; Pictish; Scottish and Latin. These languages parallel those given, in Latin, by [Bede] (1:11) in the 8th century: "Anglorum; Brettonum; Scottorum; Pictorum and Latinorum." Summing up then, there were, according to Bede and Wyntoun, just five languages. As mentioned above, one can only give a brief summary, to whet the appetite, but the ambitious reader will have no problem locating the many in depth studies, of our fascinating language history, in books (the one's mentioned above are a good starting point), journals and online resources, of which there are many. The Y chromosome proof of language distribution, in particular, is just one interesting avenue to follow. One of the problems I have discovered is that credit, for exploration and travel by ancient people, has not always been very forthcoming in the past, if Pytheas of Massalia made it to our shores, in about 325 BCE, then so could the northern Spanish Basques. On this see Oppenheimer (151, 282 and 418) and for those wishing to explore, more fully, the languages below, the Glottolog and Ethnologue online sites are two excellent resources which have comprehensive catalogues of all languages worldwide, maps and bibliographic databases.

Pictish

It is probably safe to say that Pictish was present in these islands before the arrival of the Celts. A number of etymological authorities have suggested connections with Basque, but, having said that, the Pictish language problem is a contentious, fragmentary, ongoing story of a people of which very little is known. We only have a few early mentions; one in the *Panegyrici Latini* (c CE 297-8), a collection of Roman orations, and another by Marcellinus given below. Another reference to the Picts is given in *Widsith*, an Old English poem preserved in the Exeter Book, a 10th century collection of Old English poetry. Line 79 gives; "Mid Scottum ic wæs ond mid Peohtum…", 'With Scots I was and with Picts' – two of the 70 tribes mentioned in the poem, most of which can be identified with certainty today. R. W. Chamber's published a very detailed analysis of the work in 1912, which is listed in the bibliography and there are, of course, many, many detailed accounts of the

Picts and their language available in books and online. To this day the whole issue of the Picts and their language is still vigorously debated. Needless to say, there are no known watercourse generic survivals, as the language was more than likely assimilated into Scottish Gaelic at some time, but even this statement is controversial. Adamnan (25), who wrote his *Life of Saint Columba*, in the 7th century, mentions that Columba used an interpreter to converse with the Picts, which means that the language they used was not Irish Gaelic. Bede (8th century) also mentions the Picts (1:11) as "Pictorum", one of the five indigenous groups living in Britain during his time. Prior to Adamnan's and Bede's time, we are reminded of another mention; Skene (1867:lxxxviii) states: "we learn from a passage of the Roman historian, Ammianus Marcellinus, who describes the first great outburst of the Barbaric tribes upon the Roman province in Britain, in the year 360, when he says, under the year 364 Picti Saxonesque et Scoti…", a Latin text which comes from book 26, chapter 5 of his *Res Gestae*.

Celtic

The Celts are the earliest known arrivals in the British Isles to leave a mark on our language. The arrival dates vary, often given as somewhere between 1000 and 500 BCE. Hecataeus, the Greek writer, first mentions the 'Celts', as *Keltoi*, in 517 BCE and Edward Lhuyd [Lhuyd] is usually credited with the first mention of the term 'Celt' in 1707. There are two groups of languages; the Brythonic – Breton, Cornish and Welsh, the P-Celtic and the Goidelic – Irish, Manx and Scottish Gaelic, the Q-Celtic. The P and Q Celtic synonyms and variants, 600 plus, account for more than 30% of all entries in this book, and although Breton is not spoken here, the listing of their terms is essentially given to show the relationship with Cornish, which was taken over to Brittany during the 5th and 6th century, by migrating tribes from SW Britain.

The P-group; Cornish and Welsh together with Breton, has been greatly affected. In Cornish all traces of original generic terms have disappeared from the modern map since becoming anglicized and, in Welsh too, although not as quickly, the evidence suggests that fringe areas and border incursions have resulted in the loss of indigenous terms. Breton, the branch of Cornish taken to and further developed in Brittany has also suffered, in that terms like *aven* have now been replaced by *rivière* or *stêr*. Although *aven* can still be found in compounds such as Pont-Aven, the majority of the original Cornish terms carried over are now only found in dictionaries. Breton, like its sister language, has been in decline, especially in the last 100 years, such that the people of Brittany are now having to fight hard to save their language from becoming moribund and disappearing altogether, a fate that had initially befell Cornish before its current revival. The last surviving native speaker of the Manx language died in 1974 but since then a revival effort has seen renewed interest in what would, were it not for it, be an extinct language.

The P-Celtic group have many sources – [Le Gonidec's] *Dictionnaire Breton-Français* (1850) and [Henry's] *Étymologique des termes les plus usuels du Breton Moderne* (1900) are standard Breton works to which we can add Du Rusquec's [Bret L] *Nouveau Dictionnaire Pratique et Etymologique du Dialecte De Léon* (1895) and Ernault's, [Bret V] *Dictionnaire Breton-Français du Dialecte de Vannes* (1904) for some of the dialect regions. Cornish sources are plentiful – [Pryce's] *Archaeologia Cornu-Britannica* (1790) is the oldest dictionary. [Norris] follows with his *Ancient Cornish Vocabulary* (1859); [Williams] with his *Lexicon Cornu-Britannicum* (1865) and [Jago] with his *An English-Cornish Dictionary* (1887). [Nance] *An English-Cornish and Cornish-English Dictionary* (1978) and the dictionaries of [George], *Gerlyver Kernewek Kemmyn - An Gerlyver Kres* (1998) and *An Gerlyver Meur* (2009) bring Cornish up to date. Welsh dictionaries are plentiful too. [Salesbury's] *A Dictionary in Englyshe and Welshe* (1547) is probably the earliest,

[Evans] follows, with *An English and Welsh Dictionary* (1852-8) and [Pughe & Pryse's] *Geiriadur Cenhedlaethol Cymraeg a Saesneg* (1866 & 1873) completes the older stock. However, the standard work of reference for Welsh these days is the *Geiriadur Prifysgol Cymru* [GPC], in 4 Volumes, published between 1967-2002 with a 2nd edition on the way as well as an online version of headwords and definitions, currently being enlarged to include scanned pages.

The Q-Celtic Gaelic languages, Irish, Manx and Scottish Gaelic have nowhere near the number of variant forms found in English, primarily due to the fact that their language, unlike the English, has remained free of dialectal development because the Gaelic regions have not been peopled by the same influx of nationalities that have arrived in England. Incursions into the Gaelic regions by the English have resulted in the decline of original word forms especially in southern and northern Ireland – not quite totally anglicised yet, but heading in that direction. The same can be said for the Isle of Man and parts of the Highland regions of Scotland. River has gradually replaced the abhainn's; awin's and abhuinn's to such an extent that, in the Isle of Man, for instance, the frequency of river just about exceeds every other Manx generic term that can be found on the modern map – the same goes for Northern Ireland and to a lesser extent, parts the Irish Republic, as well as Scotland.

The best sources for the Q-Celtic group are the works of [O'Brien] *Focaloir Gaoidhilge-Sax-Bhearla* (1832); [O'Reilly's] *An Irish-English Dictionary* (1864); [Dinnen's] *Foclóir Gaedhilge agus Béarla, An Irish-English Dictionary* (1904) and the *electronic Dictionary of the Irish Language*, [eDIL]; considered the standard work of reference on Old Irish. For the Scottish Gaelic, the *Dictionarium Scoto-Celticum* [DSC] compiled by The Highland Society of Scotland (1828); [Dwelly's] *The Illustrated Gaelic-English Dictionary* (1918) and [MacBain's] *An Etymological Dictionary of the Gaelic Language* (1911) are three of the best and [Kelly's] *Fockleyr Manninagh as Baarlagh* (1866) is best for Manx forms.

The Celtic P- and Q- language groups go back a long way, and as such, there is, of course, a massive amount of literature on the subject; it's really a matter of choice, depending on which particular field of research interests you, for Proto-Celtic, Matosovic's *An Etymological Lexicon of Proto-Celtic* is recommended, especially pp 2, 12, 56, 119, 147, 162, 178, 201 and 218.

Anglicized Irish

Anglicized Irish came about due to incursions into the Gaelic regions by the Normans of Wales initially, and later, by the English, and as mentioned above, these incursions resulted in the decline of original word forms especially in Northern Ireland, and to a lesser extent, in the south. Brook, burn, river, stream and water are now the common appellatives in the north and have replaced nearly all the indigenous terms, especially on maps. A typical example comes from the 17th century, when English cartographers carried out the Irish survey. The designation for the majority of Irish river terms, marked on maps, at that time, was flu and water, much the same as the terms found on English maps, first and foremost, because the surveyors did not take into account the Irish generics. Terms such as sruhan and sruthan still survive and are extant on the modern map, and are likely to stay there, but spoken and written forms are rapidly in decline.

Latin

The Romans introduced Latin into Britain in 43 CE, it was used as the language of administration, and later, the church, which was carried right through the Anglo-Saxon period and strengthened with the arrival of Pope Gregory's envoy, Saint Augustine in the 6th century. The synonyms given for Latin come from a vast number of sources of varying dates. Bede (8th century) gives us "amnis", 'river', from his *Historia Ecclesiastica* (1:332). In 1884, Thomas Wright and Richard Paul Wülcker [WW], published their 2nd Edn., of the *Anglo-Saxon and Old*

English Vocabularies containing one of the earliest Latin word lists: the *Anglo-Saxon Vocabulary 8th Century*. (1-54), which lists: *"Latex"*, the Latin and, *"burne"*, the Anglo-Saxon, meaning 'water, liquid, fluid'. [Lindsay] too, has many forms taken from the *The Corpus, Epinal, Erfurt and Leyden Glossaries* (1921a & b), and, Charles Trice [Martin] furnishes us with a wealth of terms he collected and published in his *The Record Interpreter* (1910). In addition, Old English charters, volumes of state, documents and deeds have all contributed to the Latin synonym list. There is no real difference between the uu, uv or vu, forms found in Latin words for watercourse synonyms. Charter forms, taken from Kemble and Birch [KCD] and [BCS], in particular, exhibit these forms in the flu- and ri- stem words; fluuius and fluvius – riuulo and rivulo. KCD is best for words with uu forms: "fluuius" and "riuulo" – BCS for the uv and vu forms: "fluvius and "rivulo".

English

The traditional date for the arrival of the 'English', carried by migrating tribes of Angles, Saxons and Jutes, from the Germanic countries, is 449 CE, a date which is questioned even now by a number of authorities.

It is important to note that Marcellinus, who mentions the *Picti*, (above), mentions the Saxons too. The 4th century text, in his *Res Gestae* (XXVI, 5), gives: "Picti Saxonesque et Scoti...." This could mean that the Saxons were present in Britain at this time or, merely advancing on the south east coast. The latter is favoured, as an earlier date would conflict with the traditional arrival date of 449 CE. The view that the Anglo-Saxons and other Germanic tribes were present in Britain prior to 449, is, however, gaining momentum from a number of leading authorities, especially in the field of genetics. Some controversial opinions, regarding the arrival of English, can be found on many Internet sites such as www.proto-english.org, under the title: *"How old is English."* A translation of the

Marcellinus text is available from LacusCurtius [LC].

Anglo-Saxon was the early term used; Old English was not introduced until much later. However, it soon became established as the spoken and written language in England lasting right up to the late 11th century, the time of the Norman invasion. The Anglo-Saxons gave us the first charters, written in Latin, often with English estate bounds, which correspond with modern civil parishes today. The Anglo-Saxon charters have three main sources, the most recognised being Kemble's *Codex Diplomaticus Aevi Saxonici* [KCD] and Birch's *Cartularium Saxonicum* [BCS], both now fully collated in Sawyer, *Anglo-Saxon Charters: an Annotated List and Bibliography* [S]; available electronically. For 'Old English', Bosworth's *An Anglo-Saxon Dictionary* (1898) and Toller's *An Anglo-Saxon Dictionary: Supplement* (1921) [B-T], are considered the standard works of reference; both available in electronic form from the Faculty of Arts, Charles University, Prague. English and mutual Scottish terms account for well over 1000 or more watercourse synonyms, mainly due to the growth of regional dialectal development in both languages, in fact there are more variants to be found in English and Scottish, collectively, than anywhere else in Great Britain. There are literally thousands of sources available when it comes to English, from *The First English-Latin Dictionary* (1440) to the market leader in English dictionaries: *The Oxford English Dictionary* [OED], first published between 1888 and 1928. The other major work of reference is *The English Dialect Dictionary* [EDD], based on collections of words from the English Dialect Society's own 80 volumes. The 6-volume EDD dictionary, collated by Joseph Wright, was published between 1898 and 1905.

Scottish

Many have raised the question – what is Scottish? Is it a dialect of English or a language in its own right? There are certainly plenty of shared watercourse terms to be found, as mentioned above, which have no doubt come about following English incursions into the Scottish Lowland regions, but, for the purpose of this work, Scottish words are those found in Scottish dictionaries and documents – if its in a Scottish dictionary I take it to be Scottish. *An Etymological Dictionary of the Scottish Language*, by [Jamieson], 5 Volumes, (1879-87), is brilliant. Following that we have the 12-Volume *Dictionary of The Older Scottish Tongue* [DOST] and the 10-Volume *Scottish National Dictionary* [SND], both now available online under the collective title of The *Dictionary of the Scots Language* [DSL].

Norn

The Norn language was brought to the Shetland and Orkney Islands during the 9th century by Norse settlers, and probably became extinct during the 19th or late 18th century. The Norse language is classed as an Indo-European North Germanic language. Danish, Norwegian, Swedish, Icelandic and Faroese all emanate from Old Norse and the Norn of Shetland and Orkney is another branch of it. Each island has slightly different forms. *The Orkney Norn* by [Marwick] was published in 1929; prior to that, *An Etymological Glossary of the Shetland & Orkney Dialect* by [Edmondston] came in 1866 followed by A *Glossary of the Shetland Dialect* by [Stout] in 1914. Some of these are quite small and have been overshadowed by the work of the Danish scholar Jakob [Jakobsen] – *An Etymological Dictionary of the Norn Language in Shetland*, published in 1928-32 His earlier work, *The Dialect and Place-Names of Shetland* (1897) is excellent too. One only has to look at the word 'beck' to see the relationship between Norn and some of the other Germanic languages: Danish has *bæk*; Swedish, *bäck*; the Netherlands use *beek* and the German forms are *Bach*; *Bäke*; *Beck* and *Bek(e)*. France, Luxemboug and Switzerland also use the form *bach*, and the Shetland Norn examples, collected by Jakobsen: *"bakk"*; *"bakki"*; *"bekk" and "bekki"*, are all variants of the Old Norse *'bekkr'*.

Anglo-Norman

Anglo-Norman was brought to England during the Norman Conquest in 1066. It was used as the language of the ruling classes and also in law up to the middle of the 15th century. *A Dictionary of the Norman or Old French Language* by [Kelham] was published in 1779 and [Moisy] published his *Glossaire Comparatif Anglo-Normand* in 1889. Between these dates, two major Anglo-Norman works were published – the *Liber Cusumarum* in (1860) and the *Translation of the Anglo-Norman passages in Liber Albus, Glossaries, Appendices, and Index* (1862), both edited by [Riley], are very useful Anglo-Norman sources. The advent of the internet has revolutionised the book market in recent years with the rise of more and more e-books at the expense of the print version, and dictionary projects, such as the ones mentioned above, are now commonplace. Anglo-Norman is no exception; a project by the Aberystwyth and Swansea Universities known as The Anglo-Norman On-Line Hub, now host an online version of *The Anglo-Norman Dictionary* [AND].

Romani or Gypsy

Romani is the language of the Gypsies who originated in India. It is believed that they arrived in Great Britain at the beginning of the 16th century. Unfortunately, there are very few word books devoted to the language – Gypsy words are difficult to come by – one such, is that of [Smart & Crofton] who, in 1875, published their *The Dialect of the English Gypsies*, which gives only four generic watercourse terms, one of which is *"náshin paáni.* A stream, running water". They also list *"panái, páni, or paúni"*, as "water." It is interesting to note here that *"pani"*, Old Indo-Aryan, *'paniyam'*, is found in Gujarati and Hindi, showing the antiquity of words found in Gypsy. The other terms given by Smart & Crofton are *"doriov, doyav"* and *"nill"*, all listed in the Watercourse Synonym section.

Basque
A final note on which to ponder

The Basque language problem has been gaining momentum over recent years following the rise in genetic analysis linking it to some parts of the British Isles – in 2003, the BBC, in a research programme on the Vikings, mentioned that there were strong genetic links between the Welsh; the Irish Celts and the Basques of northern Spain, so, perhaps one day, a definite connection will be made; a connection we eagerly await. The Basque language is, of course, very problematical, it had no known affinities in any of the Indo-European language groups prior to the publication of *A First Etymological Dictionary of Basque as an Indo-European Language*, by Gianfranco Forni, in 2014, and yet, in view of the foregoing, it is now gradually becoming accepted as a possible early language arrival in the British Isles, even predating Pictish! Genetics exposes the hidden history and eliminates the guesswork when it comes migrations.

Notes on the Place and River-Names

Place-Names

It is a well know fact that, of all the names extant on the modern map, mountain, river and watercourse names and terms are the ones most likely to survive, frequently taken in and carried through successive language periods, by groups of new arrivals. There are exceptions though; in the case of the river-name Avon, from the Proto-Celtic *abon (see avon in the synonym list), the incoming Celtic settlers; Irish, Manx, Scottish, Cornish and Welsh, all took over the avon term for what it was – a generic – but the Anglo-Saxons mistakenly applied it to the names of rivers, and it has stuck ever since – until recently!

There are thousands of place and river-names inextricably linked to watercourse generic terms. The 2 volume English Place-Name

Elements [EPNE] is a good starting point for those wishing to see lists of elements relating to place-names. The 85 plus volumes of The English Place-Name Society [EPNS] are a unique collection of surveys covering most of the English counties with additions, and special one-off editions, published in their *Popular Series*. The earlier, *The Concise Oxford Dictionary Of English Place-Names*, [DEPN] and the later, *The Cambridge Dictionary of English Place-Names* [EPN], both have comprehensive coverage.

For Scottish names, *Scottish Place-Names: Their Study and Significance* [Nicolaisen], which includes an in depth study of river-names; *The Dictionary of Scottish Place-Names* [Darton] and *Scottish Place-Names* [Ross], collectively, offer widespread coverage. *The Dialect and Place-Names of Shetland* [Jakobsen] and *Lewis Place-Names* [Mackenzie] are both very worthy additions.

The Celtic group has many sources, taking them alphabetically, for the Cornish, we have, *A Glossary of Cornish Names* [Bannister], *Cornish Names* [Dexter], representing the older stock; *Cornish Place-Name Elements* [CPNE], *A Concise Dictionary of Cornish Place-Names* [Weatherhill] and *Selected Cornish Place-Name Elements* [Bond] from later generations.

The Origin and History of Irish Names of Places comes from the pen of the Irish place-names master [Joyce]; *Irish Place Names* [Flanagan] is compact and wide-ranging and *The Master Book of Irish Placenames* [MBIP] is an excellent gazetteer.

For NI place-names, see www.placenamesni.org for all current and historic forms of Northern Irish place-names. It presents the material collected by the Northern Ireland Place-Name project, which was run by Dr Kay Muhr and Patrick McKay, for 23 years up to 2010, at which time it was taken over by Queens University, Belfast, who now run the site. During that period, a second edition of *A Dictionary of Ulster Place-Names* [McKay] was published covering 1300 place-names in the nine Ulster counties.

Manx place-name books are limited – *The Surnames & Place-Names of the Isle of Man* [Moore], *The Place-Names of the Isle of Man with their Origin and History* [Kneen] and *A Dictionary of Manx Place-Names* [Broderick] are three of the best available.

For Scottish Gaelic, the leader has always been *The History of The Celtic Place-Names of Scotland* [CPNS]. The *Place-Names Highlands & Islands of Scotland* [MacBain] is praiseworthy and the works of [Milne] *Celtic Place-Names in Aberdeenshire* and *Gaelic Place-Names of The Lothians* are full of questionable generic terms, most of which, remain unsolved to this day.

For Welsh place-names, the earlier, *The Place-Names of Wales* [Morgan] and the more recent *Dictionary of the Place-Names of Wales* [DPNW] are the favourites; there are many others but the two mentioned are probably the most reliable.

If Roman place-names are your interest, then the Antonine Itinerary [AI], the Ravenna Cosmography [Ravenna] and the excellent volume on *The Place-Names of Roman Britain* [PNRB] which commentates and summarises both, are hard to beat.

River-Names

When it comes to river-names, it has to be the one and only *English River-Names* [ERN] by Eilert Ekwall, published in 1928, and still considered the standard work of reference on river-names despite various challenges.

Common Etymologies for River-Names

Some river-names have a common etymology, and come from watercourse generic terms such as 'burn', 'ea' and 'water'. As [McLure] (194) points out: "Tan, possibly for an earlier Tam, of which the meaning is obscure, but from the number of river-names in which it occurs it seems to have been almost a generic word for a stream ….." This statement does have some currency, the Indo-European root *ta- is a stem with meanings such as 'to melt, to flow' and many river-names are rooted to it. [Pokorny] (1053-4), for instance, lists the Thames, the Scottish Tain and Tay under this root. So McLure had a point in 1910. The river-names Avon, Axe, Blythe, Bourne, Dover, Eye, Stour, Ray, Rea, Tame, Teme, Thames (mentioned above), Wye and many others, all have common etymologies, from various root meanings or misdivision of a term (as in Ray and Rea), but, it's not always the case; the stem We- in Weaver and Wevery, for example, is not that common, and, sure enough, no common etymology has been found. The same can be said of some other river-names.

Notes on the Sources and Bibliography

A good bibliography is an essential requirement for this type of book. Here you will find over 550 sources and bibliographic references, from Julius Caesar's, *De Bello Gallica* (100-44 BCE), to the latest, *Alt Wörterbuch* collection, published by Professor Gerhard Köbler in 2014; a compilation, not only useful for the synonyms listed in this book, but, hopefully as an aid to researchers in other fields mentioned in the preface.

County abbreviations used, are those that existed prior to the 1974 county boundary changes, on account of, in particular, the numerous volumes of the EPNS and many other standard works of reference quoted herein, which over time, have established a definite benchmark.

The vast majority of books, some 550 plus, were not available electronically at the outset of this work, such as the English Dialect Society glossaries [EDS]; Ancient Deeds [AD]; [Köbler's] mammoth Germanic collection and the massive volumes of the United States Board on Geographic Names [USBG], just to name a few. Since then, a gradual number of ever-increasing electronic resources [eres.] have been made available on the World Wide Web [WWW]. Some of the best are: Internet Archive, Google Books and Project Gutenberg. The Internet Archive have amassed what must be one of the greatest collections of English books (with search facilities) collected and scanned, by American and Canadian university libraries, that can be found under one roof! A similar collection can be found at Google Books. Virtually any book listed in the bibliography can now be obtained by pasting the title and (or) the author into the search box. In addition to the standard works of reference mentioned above, in the acknowledgements and elsewhere, other important works worth drawing attention to are listed below – just a few of the hundreds available. Square brackets indicate the coded reference, listed in the Sources and Bibliography section.

One of the most world-famous manuscripts listed in the bibliography is the *Codex Argentius* [CodexA], the "Silver Bible", preserved in the Uppsala University Library, Sweden and available online. It contains fragments of the Gospels in Gothic recorded by the 4th century Bishop Ulfilas of the Goths. It was discovered in the 16th century at a monastery in Germany and it's in this work that the remains of the East Germanic Gothic language is found, in particular, the relative watercourse terms; ahwa, 'water'; flodus, 'flood, stream' and rinno, 'brook'.

Bosworth's *The Gothic and Anglo-Saxon Gospels in parallel columns with the versions of Wycliffe and Tyndale,* published in 1865 [Bosworth], is a compilation of the four main sources of parts, in some cases, and all in others, of the Bible as a whole. John Wycliffe (1311-1384) [Wycliffe], translated his version from the Latin Vulgate, the first ever translation of the Bible into vernacular English, whilst that of William Tyndale (c1494-1536) [Tyndale], was translated from the Greek and Hebrew.

Mention must be made of two of the best English-Latin 15th

century word books; the *Promptoriun Parvulorum Sive Clericorum* [PP], dated 1440, the *first* English-Latin Dictionary and the *Catholicon Anglicanum*, [CA], dated 1483. A Catholicon is a comprehensive treatise usually published in a dictionary format, such as the *Catholicon* mentioned above. The French *Catholicon* is a Latin-Breton-French dictionary compiled in 1464 by a priest of Tréguier called Jehan Lagadeuc and published in 1499; it was the first printed French dictionary; the first ever trilingual dictionary and the first ever book to be published in Breton.

Archaeologia Britannica by [Lhuyd], published in 1707, is a masterly work for all things British; languages, histories and a large comparative Celtic etymological section, considered the first significant work on the subject; an Irish-English dictionary, and, an in depth study of the Cornish language. He must surely be one of the greatest antiquaries to have ever lived.

The dialect words collected by Joseph Wright in *The English Dialect Dictionary* EDD, 6 Volumes, published between 1898 and 1905 are collections of words from the English Dialect Society's [EDS] 80 volumes, published between 1873, when the society was founded by the Reverend Professor W. W. Skeat, until its closure in 1896. The EDD is a really massive collection of all things dialect; the greatest work on the subject to be found anywhere.

All former Middle English dictionaries have now been superseded by the Middle English Dictionary [MED], published by the University of Michigan Press, 2003, when the last of its 115 Fascicles were published, 75 years after its commencement. It has been described as "the greatest achievement in medieval scholarship in America." There are 15,000 pages; over 7,000 main entries; a massive corpus of Middle English texts and a colossal bibliography. So, if ME is your subject, then look no further than the MED.

Jamieson's *An Etymological Dictionary of the Scottish Language* [Jamieson], 5 Vols., published in 1879-87, is a must for all words Scottish and [King's], Ph.D. thesis, *Analytical Tools for Toponymy: their Application to Scottish Hydronymy*, is now the most comprehensive work of its kind on the subject.

The Shetland Norn words collected by the Danish scholar Jakob Jakobsen, published in two volumes, as *An Etymological Dictionary of the Norn Language in Shetland* [Jakobsen 1928], is still *the* very best standard work of reference on the Norn language to this day. 10,000 words, initially planted by Norse invaders, would probably have been lost had it not been for such foresight on his part.

For Indo-European synonyms, Carl Darling Buck's [Buck], *A Dictionary of Selected Synonyms in the Principal Indo-European Languages*, is hard to beat. A fantastic, compact, yet comprehensive listing, of not just watercourse terms, but all things Indo-European, in 1000 synonym groups.

Dwelly's *The Illustrated Gaelic-English Dictionary;* Pokorny's *Indogermanisches Etymologisches Wörterbuch*; now superceded by the massive Germanic collection of Gerhard Köbler and Lewis and Short's [L & S] *A Latin Dictionary*, are just a few more commendable works.

Notes on the Maps

Historically, early maps covering the British Isles, such as the 14th century Richard Gough Map, are quite late compared with the ones ancient civilisations made on clay tablets, thousands of years ago. The oldest known map ever found, showing canals, streams and irrigation channels comes from the ancient city of Nippur in Mesopotamia, the modern day Iraq, a clay tablet with a cuneiform inscription dated 14th – 13th century BCE.

The online Ordnance Survey [OS] get-a-map service enables speedy checking of any given map reference. Supporting the OS range of maps is the [Geograph] site. A British and Irish project which aims to collect geographically representative photographs and information for *every square kilometre* of Great Britain and Ireland, using the National Grid Reference [NGR]; an excellent resource for viewing watercourses nationwide. In addition, the OS Northern Ireland Discoverer Series [OSNIDS] and OS Ireland Discovery Series [OSIDS] maps are available in paper form as is the Isle of Man survey 1:25000 series, published in two sheets. The [USBG] is a *must* and covers all countries worldwide – place-names, river-names and all generic terms.

According to Geograph there are 340,000 grid squares on the OS Explorer series maps and I spent five years looking at them, together with other maps, a few at a time each week. Gradually, all 470 OS Explorer series maps [OSX] were checked, a square at a time, and yet, I am always finding myself going back there, only to find that I had missed something! The Irish, Manx, Scottish and Welsh maps were treated likewise, and, in addition, hundreds of old county and regional maps, such as the 19th century, OS First Series maps had to be looked at too, the grand total being somewhere in the region of a good thousand or more. It was certainly worth doing because it threw up gems such as the Old English '*burna*', still extant on the modern map, after its first mention over a thousand years ago; 'scurth', an old form of 'skirth', from the Old Norse '*skurðr*'; 'sick', 'sike' and 'syke', three distinct forms from the Old English '*sic*'; a Scottish '*den*'; a mystery Scottish Gaelic '*allt*' in Wales and a Scottish Gaelic '*gro*', commonly found on the Isle of Lewis. An anglicized Irish '*owen*' (their anglicized form of avon) alongside Irish '*abhainn*'; a Manx '*awin*' and a probable Latin survival in '*foss*'. In total, there are over 400 mapped generic term examples marked on the modern map covering the area surveyed!

Old Maps

The British Museum has a fantastic range of old maps covering Great Britain, viewable in their Online Gallery, and, the National Library of Scotland has a similar range too, including lots of old OS maps. Old English, Scottish and Welsh county maps are available from Genmaps; Genuki has hundreds of links to old maps countrywide including Ireland and *The Master Book of Irish Placenames* includes a complete set of 19th century maps for every county. In addition, a searchable database for all OS maps is hosted by old-maps.co.uk.

ABBREVIATIONS AND PRINCIPAL CONTRACTIONS

The conventional recognised standard abbreviations and principal contractions are used for country, county, language and other miscellaneous entries with slight modification where conflicts occur within these standards.

* Hypothetical or postulated form

AIr.	Anglicised Irish	Ed.	Editor
Alb.	Albanian	Edn.	Edition
A-N	Anglo-Norman	Eds.	Editors
A-S	Anglo-Saxon	Egyp.	Egyptian
Bas.	Basque	Eng.	England
Beds.	Bedfordshire	Er.	Erse
Bel.	Belgium	eres.	electronic resource
Berks.	Berkshire		
biblio.	bibliography	Ess.	Essex
BM	British Museum	ety.	etymology
Bret.	Breton	F	French
Brit.	British	Fa.	Faroese
Bucks.	Buckinghamshire	FI	Faroe Islands
C	Century	Fle.	Flemish
c	circa	FN	field name
Cambs.	Cambridgeshire	Fra.	France
Celt.	Celtic	Fris.	Frisian
Ches.	Cheshire	fwd.	foreword
Co.	Cornish	G	German
cog.	cognate	Gael.	Gaelic
Corn.	Cornwall	Gaul.	Gaulish
Cumb.	Cumberland	GDR	German Democratic Republic
Da.	Danish		
Den.	Denmark		
Derbys.	Derbyshire	GE	German Etymology
Dev.	Devon		
dim(s).	diminutive(s)	GFR	German Federal Republic
Dor.	Dorset		
Dur.	Durham	Gloucs.	Gloucestershire
Dut.	Dutch	Gm.	Germanic
E	English	Gmy.	Germany

Got.	Gothic	Ms.	Manuscript
Gyp.	Gypsy	Mss.	Manuscripts
Hants.	Hampshire	Mx.	Manx
Herefs.	Herefordshire	N & Q	Notes & Queries
Herts.	Hertfordshire	N	Norse
Hitt.	Hittite	NE	New English
Hunts.	Huntingdonshire	NGR	National Grid Reference
Ic.	Icelandic		
Ice.	Iceland	NHG	New High German
IE	Indo-European		
IG	Indo-Germanic	NI	Northern Ireland
IGael.	Irish Gaelic	Nn.	Norn
IOM	Isle of Man	Nor.	Norwegian
IOW	Isle of Wight	Norf.	Norfolk
Ir.	Irish	Northants.	Northampton-shire
Irl.	Ireland		
Ital.	Italian	Notts.	Nottinghamshire
L	Latin	NPr.	New Prussian
Lancs.	Lancashire	Nrth.	Northumberland
lang(s.)	Language/s	Nth.	Netherlands
Leics.	Leicestershire	Nwy.	Norway
Lett.	Lettish	O	Old
LG	Low German	OBrit.	Old British
Lincs.	Lincolnshire	OCelt.	Old Celtic
Lith.	Lithuanian	OE	Old English
LL	Low Latin	OF	Old French
Luw.	Luwian	OFris.	Old Frisian
Lux.	Luxembourg or Luxembourgeois	OGael.	Old Gaelic
		OGaul.	Old Gaulish
M	Middle	OHG	Old High German
MDut.	Middle Dutch		
ME	Middle English	OIc.	Old Icelandic
MHG	Middle High German	OLFrk.	Old Low Frankish
Middx.	Middlesex	ON	Old Norse
ML	Medieval Latin	onl.	online
ModE	Modern English	OPr.	Old Prussian
ModL	Modern Latin	OPr.EV	Elbing Vocabulary
ModSc.	Modern Scottish		

OrkNn.	Orkney Norn	Som.	Somerset
OS	Ordnance Survey	Sp.	Spanish
OSax.	Old Saxon	Sr.	Surrey
OSc.	Old Scottish	Staffs.	Staffordshire
Oxon.	Oxfordshire	Su.	Sudovian
p	page	Suf.	Suffolk
PC	Proto-Celtic	Sum.	Sumerian
Pg.	Portuguese	Supp.	Supplement
PIE	Proto Indo-European	Sw.	Swedish
		Swe.	Sweden
PN	Place-Name	Switz.	Switzerland
pp	pages	Sx.	Sussex
R	River	USA	United States of America
Rads.	Radnorshire		
Rev.	Reverend	Vol(s).	Volume(s)
RN	River Name	Wal.	Wales
Russ.	Russia	Warks.	Warwickshire
Rut.	Rutland	Wel.	Welsh
S & O	Shetland & Orkney	West.	Westmorland
		Wilts.	Wiltshire
Sax.	Saxon	Worcs.	Worcestershire
Sc.	Scottish	WWW	World Wide Web
Scot.	Scotland	YE	Yorkshire East Riding
SGael.	Scottish Gaelic		
Shet.	Shetlandic	YN	Yorkshire North Riding
Shet. Isl.	Shetland Islands		
Shrops.	Shropshire	Yorks.	Yorkshire
Skt.	Sanskrit	YW	Yorkshire West Riding
SNn.	Shetland Norn		

WATERCOURSE SYNONYMS

a (Dut., Fris., OFris & Sum.) From the ON *á*; probably the most common element to be found with a 'river, stream' meaning, surviving, as a prefix or suffix, in many PN's and RN's, found in most of the northern and eastern counties of Eng., Scot., the IOM and the Shet. Isl. associated with ON settlements. Aby in Lincs. and the Yks. Ayresome, Ayton and Greta, just to name a few, are all linked to this widespread element. The terminal o in Thurso, Scot., is from the same root. Broderick (197) has Mx. survivals such as Laxey, formerly Laxá (Moore 284), identical with Laxá, 'salmon river', in Ice. and Ellwood (1895:1) states that: "In Edda over one hundred North-English and Scottish rivers or a's are mentioned." OFris. *a* is listed by Köbler and Halbertsma (1) gives: "*A*" as Fris. Jamieson (Supp., 5:1a) gives *a* as a Sc. term denoting: "Water; and applied in various ways to the sea, a river, stream, spring, fountain, & c., of which there are abundant traces remaining in almost all the districts colonised by Norsemen or Danes". However, he does not give any particular instance of its use as a generic term. There is a lengthy discussion on *a* by Embleton (1-4). and EPNE (1.1); ERN (1); Köbler (ON) and all the EPNS volumes associated with the Norse regions of the British Isles offer, collectively, a colossal amount of research material on this element. The PNYW (7:286-7) list over one hundred elements associated with watercourse terms; *á*; is the first. As a generic term, the *a* form is used in the Nth; the ON *á* in the FI and Ice. and *å* in Den. and Swe. Another interesting fact about a, is that in Sumerian, a language which existed about 3000 to 1600 BCE and beyond, it meant watercourse! The term is given by Waddell (2a) – Prince (3) associates it with "water" and "irrigation". Both confirmed by Halloran (3). See aa; ae; aw; ea; and o; The IED (38b); Waddell (2a); Halloran (3) and the appropriate USBG entries.

á (Fa., Ic. & ON) See a and aa.

å (Da. & Sw.) See a.

aa (Da., Du., E, Fle., G & Sc.) An obsolete term for a stream or watercourse, first recorded in 1430 but otherwise uncommon in antiquity. The MED list this form under *a*. Macray (143) gives: "water-course or drain", with a Swe. and Da. etymology, cog. with OE *éa*. For the Sc., Jamieson (1.1a) lists aa, as one of five two lettered words, all defined as: "applied in various ways to the sea, a river, stream." Although aa is obsolete in Eng., it is still used as a generic term in Bel., USBG (Bel.:iiib); Gmy., USBG (GFR:iiia) and the Nth., USBG (Nth. fwd:a), in the sense 'stream'. In addition to *å*, Nwy. uses the form *åa* as a generic stream or river suffix in Krosså and Juråa etc., Den. also uses *aa* as a common suffix in the names of rivers and streams and Fra. has a few too. In addition, aa, as a RN, can also be found in Latvia (as an old name of rivers); Switzerland and Ethiopia. All the terms are cognate with the E ae, ea and ee, and well evidenced in the West Germanic branch of the IE languages.

åa (Nor.) See aa.

Aach (G) See ach.

ab (OIr.) The stem of all Irish watercourse generic forms beginning with ab; i.e., aba, abh, abha, abhainn etc., eDIL lists over twenty different forms under the headword ab. Meyer (4) and Hessens (177b) give ab, 'river'. All river words found in Ir., Mx., and SGael., formed from the stems ab-; am-; au-; av-; aw-; ob-; ot-; ou- and ow- are cognate and go back to the PIE root *ab-*. Over a period of time, this root has led to the diversity of forms we see today. See avon for the ab- stem etymology and CPNS which has many ab- forms, all listed in Index (E), as well as all the other stems listed above. Hogan's ab- list (1a-3b) is worth consulting too.

*ab (PIE) See ab and avon.

aba (Ir.) One of many ab- stem words meaning 'river, stream' listed by Hessens (177b) and Meyer (4). Adamnan (132) in his *Life of Saint Columba*, 8th C, (c713, extant MS.) gives: "Abae fluminis", one of the oldest records for the OIr. ab- stem. It can also be found in the ALI (4:144.1).

Abae (OIr.) See aba.

abainn (Ir.) See abhainn.

abann (Ir.) Two German etymologists list this form – Hessens (177b) and Windisch (342b) as "fluss, river", abhann, the more common form, can be found in eDIL.

Abbona (Wel.) The Wel. form of abona. See avon.

Abbone (OBrit) See avon.

abe (Wel.) A contraction of aber.

aber (Wel.) Most people think of this word as meaning 'mouth of river, estuary' – Aberystwyth, mouth of the R Ystwyth – Aberafan, mouth of the R Afan. However, a watercourse definition exists too! The LL, 12th C, (214) gives: "*aper*" for aber. Salesbury, in his *Englyshe and Welshe* dictionary, published in 1547, gives: "abe" [sic], a contraction of aber and Pughe and Spurrel's dictionaries, as well as the GPC, all give 'brook, stream' or similar, and, there is good geographical evidence to support this – at least four instances are shown on maps – OSX 178 at SN4709; OSX 215 at SH7606; OSX 256 at SJ3046 and OSX 264 at SJ0057.

aberig (Wel.) A dim. of aber, 'small stream, rivulet', Evans (2:711a) and Pughe & Pryse (1:50b) both list it.

abh (Ir. & SGael.) According to Joyce (1869:1:454) "The Irish language has two principal words for a river – abh or abha [aw or ow] and abhainn." O'Reilly (7a) lists abh and Joyce (above) has further examples. Although abh is shown on OSIDS 45 at M0144, it is only used as a contracted form of abhainn. In SGael., it is not usually considered as an independent generic term, Dwelly (2a) gives abh as an obsolete word for water; the same is in M'Alpine and, CPNS (477), expands the term, however, abh does underlie a number of RN's and PN's in Scotland. Milne (1912:22), for instance, under the PN Auburn, states: "Both parts mean water. Abh, water, stream; burn, flowing water", but, in spite of this, contemporary toponymists usually give abh as meaning 'water' or as an obsolete word for it. See eDIL.

abha (Ir.) A variant of abh, Dineen (1904:2a) lists this form in punctum delens format, see aman. As mentioned above, under abh, Joyce states that it is one of "two principal words" used for river. O'Donovan (1856:154) gives: "River Sele or Abha-ùhubh" and, the map of County Kerry, shows two instances on OSIDS 70 at Q4303 and Q5408. See abh.

abhainn (Ir. & SGael.) One of the most common generics used in Ir. and SGael., more so than abh, to denote 'river' – usually applied to rivers of some importance as apposed to lesser streams. Joyce (1869:1:454-5) equates it with the Sanscrit *avani*. Some Sanskrit dictionaries such as Williams (1872:91a) list this under *avana*. Nennius (214) uses the form *abainn*. See OSIDS 45 at M0844 for an Irish example and OSX 355 at NR4572 for the SGael.

abhan(n) (Ir.) O'Reilly (7b) gives: "abhan, a river". The abhann form, but not abhan, has an entry in eDIL and O'Donovan (1856) gives: "the mouth of the River Abhann-mor."

abhin (Celt.) MacLean (127) gives: "The Celtic reader is aware that Abhin, or Avin; rapidly, Ain, or Oin, is the term for a river."

abhna(i)g (SGael.) Dwelly (3b) defines abhnag and abhnaig as: "a little river."

abhoinn (Celt.) MacLean (47) gives: "river."

abh-shruth (SGael.) One of the few recorded Gael. compounds. DSC (4a) defines it: "a current or rivulet."

abhuinn (SGael.) This is a rare variant of abhainn. DSC (6a) and McAlpine (2a) list it as "river". Surprisingly, it can still be found on the modern map at NGR NG5329. See CPNS (45).

*abon (PC) See avon.

Abon(a) (OBrit) See avon.

Abone (OBrit) See avon.

ach (SGael. & Wel.) This is given by Milne (1906:127a) as an obsolete word for "water, stream." For the Wel., Lhyud (6a) gives: "aches, a river." However, the GPC (7b) only defines aches as: "flood." Morgan (10) states that: "Ach is a Celtic derivative particle denoting water", and (54) "Ach, a stem, a pedigree, a river." A term confirmed by DPNW (xxiv). In

Gmy., Ach(e) is still used of a stream generic in addition to Aach and Achen – Friedberger Ach and Irschinger Ach are just two examples. See USBG (GFR:iiia).

Ach(e) (G) See ach.

Achen (G) See ach.

aches (Wel.) See ach.

ade (E) See aid.

ader (Fl.) See ǽd(d)re.

adere (Dut.) See ǽd(d)re.

ae (E, ME, OE, & Sc.) From the OE *á*, B-T (8). A river term found in the *Ormulum*, line 7091; Layamon's *Brut*, line 1400: "*in are fwiðe feire æ*"; Thorpe's *Psalmorum* (83) and in the *Supplement to Jamieson's Scottish Dictionary* (5.1a). In addition, Earle, in *Two of the Saxon Chronicles* (1865:31), from 656AD, gives: "*Bradan æ*", and the modern map, at NGR NY0485, shows Water of Ae, a river in Dumfries and Galloway, confirming a Scottish Lowland connection. Oldfield (60) also mentions this term, from an inquisition, dated 1560, given by Sir William Dugdale. The word is a variant of OE *ea* and *e* 'river' and cog. with some words in the West Germanic branch of the IE languages, namely OFris. *a, e*; ON *á*; OSax. and OHG *aha*, as well as the East Germanic Got. *ahwa* – L aqua, 'water'. The Gm. form is **ahwo* and the IG is *akwa*. (all from Köbler). See aa; ea; ee; B-T; the MED; Fick, Falk & Torp; Köbler and Pokorny for further definitions and Indo-Germanic roots.

ǽ (OE) See ae.

æa (OE) BCS1005 gives this form as a variant of *ea*. See PNBrk (3:715).

ǽd(d)re (OE) B-T (9) give: "ǽdre, ǽddre, édre, ... *an artery, a vein, fountain, river*", a term cognate with OFris, *addre*, OHG *adra* and some other Gm. languages. Kiliani (3b), for the Dut. gives: "*adere*" in the same sense. Mutschmann (1913:4) suggests that Averham, in Notts., is rooted to ǽdre but this is ruled out by Watts (2010:27). A short stream flows through Edderside, Cumbria, shown at NGR NY1045 but not named. PNCu (2:296) suggests that this is the stream that gave name to the place but, further evidence is lacking. The RN Etherow, in Ches., belongs here too and is discussed in PNCh (1:23). In Belgium, ader is still used of a drainage ditch, see USBG (Bel.:iiib); PNDb (3:679); EPNE (1:3 & 2:147); PNYW (7:182) and, on the Sc. RN suffix -adder, Nicolaisen (236-9).

affluent (E) A stream flowing into a larger stream or river, a tributary. Phillips (1853:104) states: "As an arm of the sea we shall treat of it hereafter, and now proceed to the only remaining affluent of importance on its northern banks, viz. the river Hull." Affluent, like consequent, obsequent and subsequent, are terms used more often in geomorphology rather than as a general watercourse qualifier. I have seen no map, modern or otherwise exhibiting the term.

afon (Wel.) First recorded in the 13th C., but, as with the E river; the Wel. equivalent needs little introduction, being so well known throughout the country and its borderland – it's the most common generic term applied to rivers in Wal. Salesbury, in his *Englyshe and Welshe* dictionary, published in 1547, gives: "abe [sic] ne afon A ryuer". In Wel., afon does not have the great number of variant spellings found in its E parallel – however, it still goes back to the OBrit. *Abona* and has a much rarer form, *Abbona*, attested in the LL (326 & 331). See avon.

afon fechan (Wel.) Evans (1858:711a) lists this dim. compound as: "a little river" and "riveret".

afongainc (Wel.) This is listed in the GPC (102c, 2nd Ed., 2003), explained as: "Rhagafon: tributary." its use dating from 1838.

afonig (Wel.) The dim. of afon: 'a small river, riveret or rivulet'. Evans, Pughe & Pryse and the GPC all list it and Lhuyd (165a) spells it "avonig."

afonol (Wel?) Richards (12b) gives: "little river", but the common definition of afonol, is adjectival: 'of a river' or 'belonging to a river'.

aghlish? (Mx.) Only in Kelly (1866:5a) defined as: "arm or branch of a river." The term does normally signify an arm, but not in the sense, arm or branch of a river, therefore, pending further research, aghlish remains doubtful as a qualifier.

aha (OHG & OSax.) See ae.

ahva (Got.) The original form of the Got. variant *ahwa*, 'water, river, stream'.

ahwa (Got.) A variant of the Got., *ahva* given in OED1 (143c).

*ahwo (Gm.) The Gm. form of the Got. *ahva*. See ae and *akwa.

aid (E) A special term found only in Shropshire, perhaps used as, 'an aid to draining'. Wright (1880:41b) states: "in Shropshire, a deep gutter cut across ploughed land, as well as a reach in the river, are so called." Jackson (5) defines it as: "a gutter cut across the 'buts' of ploughed lands to carry off the water from the 'reans.' – Church Stretton; Clee Hills." Under ade and aid, Hartshorne (302) applies the gutter definition to both words as well as "a reach in a river." The word 'ade' is not applied as a draining term by Jackson (3) who gives: "a reach in the Severn." The foregoing points to the fact that there has been some confusion in the past as to which of the two words define the drainage term, but, the evidence here, particularly from Jackson, suggests 'aid' is the correct one. If there are any grounds for connecting ade or aid with the OE word *ǽddre*, suggested by Hartshorne, then the link has long been lost.

ain (Wel?) See an.

*akwa (IG) The IG form of the Got. *ahva*, Balg (7b). See ae; aqua and Nostratic (200:122) for all things water.

aldie (SGael?) Only in Milne (1912:3) as: "small burn." Milne has quite a number of generic terms listed but the majority are debatable and not verified by contemporary scholarship. Liddall (1) gives: "Aldie. Allt = burn", which could have been taken from Milne. There is an Aldie Burn, near to Tain, in Ross and Cromarty, shown as Aldie Water on the latest map of the area.

aleur (F) See alure.

alien water (E) Ogilvie (1.70b) lists this unusual term stating that it refers to: "any stream of water carried across an irrigated field or meadow, but which is not employed in the process of irrigation."

all (SGael?) Only in Milne (1912:3) as: "river." A contracted form of allan perhaps, but, as with so many of Milne's terms, no other authority seems to support the fact that 'all' exists as a generic term. See allan.

allach (SGael.) According to Milne (1912:3), allach means "burn, stream" as does allachan, the dim. of allach, "small burn". He also lists "allachy, little burn." Macdonald (7) under Allachrowan, gives the same. As mentioned above, some of Milne's generic terms are debatable, however, in this case, map evidence would seem to confirm the definition, as OSX 395 at NO5291 & 5391 show two allach's flowing into the Burn of Kalfrush. In addition, the Water of Allachy is shown at NGR NO4792, and, the East and West Grains of Allachy at NO4887. See grain.

allachan (SGael.) See allach.

allachy (SGael.) See allach.

allan? (SGael.) Only in Milne (1912:3) as: "stream." If 'all' *was* the parent word then allan would be the dim. form, but, there is no evidence to support such a derivation. There is, of course, extensive literature on the name Allan, Allen and all other associated RN's, suffice to say here that the Alauna (Rav. c700) and Alaunus (Ptol. c150) derivations, favoured by early commentators on the subject, has now been firmly established by present scholarship, in particular, CPNS (467-8); ERN (4-11); Mills (10); Nicolaisen (239-40); PNRB (243-247) and Watts (8b & 10a). There are, of course, many other useful references to be found in PN books and such, so recourse to the appropriate volumes of the EPNS would be a good starting point for further in depth analysis of this term, still questionable as a watercourse generic.

alldan (SGael.) DSC (2.469b) under drill, gives: "a small dribbling brook: caochan, alldan." See alltan.

allt (SGael. & Wel.) In SGael. allt is the most common form of all generic terms used in the Gaelic regions of Scotland, defined as 'a mountain stream'. One only has to look on the modern map to see just how dense and widespread allt really is. Its use equates with the E generic 'brook', the most common form used south of the border. Dwelly (26a) and many, many others all list this term and CPNS has an excellent list of variants all listed in Index (E). It originally meant 'height, hill etc' and only later was it applied to swift mountain streams. It roots back to the L *altus* – 'height', found as a hydronym in the ancient Illyrian language. In Wel., allt is not defined as a watercourse generic, in fact, anything but, however, surprisingly, there is one marked on OSX177 at SN1908, so a rare occurrence indeed. For the Mx., Kelly (1866:6b) gives alt: "a brook or stream", which survives in PN's – see Broderick (2006:2a). Kelly's other form (13a) is: "ault, a rivulet or bourn", a variant of alt. It is interesting to note that in Sc. PN's, in particular, and some stream names, the Mx. form *alt*, as well as the SGael. *auld*, is used as a first element. See uillt.

alltan (SGael.) The dim. of allt and alt 'little brook, streamlet'. Dwelly (26a) covers the SGael., "alltan" supported by a mapped example shown at NGR NR6110. For the Mx., Kelly (1866:6b) gives: "altan, a small brook or stream" and (145a) the variant "oltan, a brook."

alltanan (SGael.) Another dim. of allt, Dwelly (26a) gives: "little brook, streamlet." See OSX 355 at NR4673 and OSX 440 at NC3532 for two mapped examples.

alour (A-N) See alure.

alt (Mx.) See allt.

altan (Mx.) See alltan.

alueum (L) WW (1:345.22) give OE: "streamrace". See alveus.

alueus (L) This is equated with 'stream, streamraad and streamracu', in the lists of WW (1:325.36; 5.10 and 178.5 respectively) from *Glosses, Latin and Anglo-Saxon of the 11th Century*. See alveus.

alura (L) See alure.

alure (A-N & E?) From the F *aleur*. In his *Dictionary of Obsolete and Provincial English*, Wright (1880:60b) gives: "alure and alour (A-N) a channel behind the battlements, which served to carry off the rain-water." Moisy (44) lists alure, defined as an alley or passage, not as a water channel and Martin (184b) gives: "alura

(Fr. aleur)" and defines it, amongst other things, "a gutter."

alveo (Ital. & L) See alveus.

alveus (E?, L) Although given in Ogilvie's dictionary (1.82c) as: "bed or channel of a river", alveus is really a L word. The Roman poet Vergil, in his epic poem *Aeneid* (7:33), and again in *Georgics* (1:203), gives: "*fluminis alveo*" and "*alueus amni*" respectively. In Italian, alveo, is still used as: 'a bed of a river'. It is equated with "stream, streamraad and streamracu", in the lists of WW (1:325.36; 5.10 and 178.5 respectively), who also give (345.22): "*alueum*, streamrace" from *Glosses, Latin and Anglo-Saxon of the 11th Century*. See alueus.

aly (SGael?) Liddall (2) gives: "Aly (a small stream)." No other authority seems to list the term.

amain (Ir.) See aman.

amair (SGael.) Dwelly (28b) gives this, and amar, the relative of the Mx. ammair, defined as: "trough, channel, ditch." The Allt an Amair shown at NGR NJ1343 does, of course, carry its own qualifying term.

amair-uisge (SGael.) This compound is given by Dwelly (28b) as: "aqueduct." See amair and amaruisge.

aman (Ir. & SGael?) A feature of Old Irish was the punctum delens [PD], a dot over a consonant, usually b or m etc., to soften the sounds between vowels and, especially, to indicate that the consonant should be followed by the letter h. Therefore, aban and aman, in PD format, should be written abhan and amhan respectively. This, apparently, has not always been the case. It is probable, in some instances, such as the entry in DSC (1:42a) that aman has been mistakenly written instead of amhan, thus presenting us with what would appear to be a new form of a word for river. CR (10:282) states: "Aman is decidedly not a parallel form of Avon, the PC form of which would be **Abona*, while that of Aman would be **ambona*." On this statement see CPNS (430). O'Brien, in his Irish-English Dictionary (1832), gives aman only, whereas in O'Reilly's Irish-English Dictionary (1864), we find aman followed by its orthographic transcription 'amhan', which makes interpretation much clearer. Hogan (30b) gives the form: "amain". CPNS (369 & 430) discusses the RN Aman in Scotland and there is another River Aman in Wal. See OSX 166 at SO0100 and amon.

amar (SGael.) See amair.

amar-uisge (SGael.) Dwelly (28b) gives "aqueduct." See amair and amairuisge.

ambe (Gaul.) See avon.

*ambona (PC) See aman.

amhain(n) (Ir. & SGael.) Another abhainn variant given by O'Reilly (27b). It has no entry in edIL, probably because it is considered a misspelling of abhainn. Hogan (51b) lists amhainn under "ath abhainn". It is found in SGael. too, with double n ending, DSC (1:43b), MacBain (1911:13) and CPNS 430 all list it, however, the MacBain and CPNS entries throw doubt on its validity. Despite this, at least four are marked on maps – one of which can be seen on OSX 446 at NC2058 confirming amhainn as a qualifier.

amhan(n) (Ir.) Lhyud (317a) and O'Reilly (27b) both give amhan as one of three am- stem variants. Hogan (30b) lists: "amhann" for some Sc. RN'S. See abhan.

amhuinn (SGael.) A variant of amhainn given in DSC (45a), related in much the same as abhuinn is to abhainn. This term can be seen, north-west of Stornoway, in the County of Ross and Cromarty, on OSMS – on modern maps, amhuinn has been replaced by abhainn! Maxwell (7 & 184) gives the form: "amuin".

ammair (Mx.) This and ammair-ny-hawiney are given by Kelly (1866:7a) as: "A channel, bed of a river, trough." Cregeen (19a) gives: "am'myr, a canal, or channel of water." The term does not appear to feature in PN's but could be related in some way to the Ir. or SGael. am- forms although confirmatory evidence is lacking. See amair.

ammair-ny-hawiney (Mx.) See ammair.

amnis (L) Bede uses amnis in his *Historia Ecclesiastica* (1:332) and KCD 698 (3:301), (S 891), gives: "amnis Auenae", the 'R

Avon'. WW (1:325.32), give the L and OE forms: "*Amnis, ea*", from an *Anglo-Saxon Vocabulary of the 11th Century*, and (546.17), "*(Amni)s*, eaa"; both 'river', from a *Semi-Saxon Vocabulary of the 12th C.*

amon (SGael? & Wel?) According to the DSC (1:42a) amon is related to aman and amhainn. Williams (1865:15b) records it as an old and obsolete Wel., word; Lhuyd (284b) also as obsolete and Skene (179) gives: "amnis amon", 'River Avon'. So, to summarize, amon as a generic term, would appear to be questionable in both SGael. and Wel. See aman.

amuin (Sc.) See amhuinn.

an (Ir.?, SGael. & Wel?) Dwelly's definition (30b) is: "an obsolete word for water." In Gaelic today, *an* means '*the*'. It may have survived in *An Iola*, marked on OSX 376 at NM9747, which would mean 'The Water of Iola'. The alternative derivation is 'The Iola'. Pennan in Grampian, formerly Aberdeenshire, is considered by some (Darton and Ross) to be 'head' + 'water', but there is no definite agreement overall. Johnston (197) states: "The only Pen- north of Perth." For the Wel., Morgan (86) gives: "an or ain, brook, signifying the running stream", an, as a terminal in Wel. is normally used as a dim. suffix, as in ffrydan and the compounds ffrwd fechan and afonig fechan – 'little river'. In Ir., an is appended to some RN's, OSIDS 35 at N7996 shows River Lagan, the anglicized form and An Largan, the Ir. form; The same map, at N8186, shows River Dee, An Nith. These terms can be misleading because the Ir. generic given for the R Lagan is very often Abhainn, not an!

aoidh (SGael.) See ùidh.

aon (Co. & Mx.) All Cornish forms from aon to awan are variants of auon or avon. Bannister (1871:2a) uses aon for auon, as does Taylor (196 & 214), but Lhyud (290a) lists it as Mx! See avon.

*ap (PIE) See avon.

apas (Skt.) See avon.

ape (OPr. & Su.) See avon.

aper (Wel.) The LL form of aber.

apis (Su.) See avon.

aqua (L & E) The L word for water, cognate with Got., *ahva*, 'water, river, stream'; commonly used as a prefix in compounds; occasionally used of a stream in OE law, confirmed by Black (82b) who states that: "In the civil and old English law. Water; sometimes a stream or water-course."

aquaductile (L) WW (1:564.47) give, from a *Latin and English Vocabulary* of *the 15th Century*, the ME: *"condyt"*. The variant *aqueductile*, (733.39) is given from a *Nominale of the 15th Century* as ME: "guttur"; E 'conduit, gutter'.

aquaductum (L) Listed by WW (1:733.37), from a *Nominale* of *the 15th Century*, which gives ME: "guttar". The variant aqueductum, (339.4) is given from the *Glosses, Latin and Anglo-Saxon of the 11th Century* as ME: "wætergelada"; E 'aqueduct'.

aquaduklyd (ME) See aqueduct.

aquaeductus (L) WW (1:184.12), from the *Supplement to Alfric's Vocabulary* (c 955-1010), give the OE equivalents: "wæterscipes", and, (1:191.5) "wæterþeote". The variant aqueductus, (1:733.40) is given from a *Nominale of the 15th Century* as ME: "cundyth undyr the erthe"; E 'aqueduct'.

aquage (E & L) This term first appears in the *Dictionarium Anglo-Britannicum*, 1708, by John Kersey under aquagium: "an aquage or water-course." Phillips (1720) and Bailey (1735) have similar entries. WW (1.118 line 12), in a list of words from Archbishop *Alfric's vocabulary of the 10th century*, give: *"aquagium"* which translates "wæterþeote." However, *wæterþeote* itself, elsewhere in the WW list, is given with a number of meanings: "aquaeductus, canalis and colimbus." L & S (148b), under *aquaeductus* and *aquagium*, translate both words: "aqueduct." Black (449b) mentions that: "EWAGE. In old English law. Toll paid for water passage, the same as aquage." See aquagium.

aquagium (E? L) This term is listed in the PP (196) as the definition of "Gote, or water schedellys", and in the *Dictionarium Anglo-Britannicum*, 1708, by John Kersey: "an aquage or water-course." It is also given by WW (1:118.11) from *Abbot Alfric's Vocabulary* together with the OE "wæterþeote" and (1:564.48), the ME: "gutur or condyt" from the *Latin and English Vocabulary of the 15th Century*. Martin (188a) gives: "watercourse" and L & S (148b) translate aquagium: "aqueduct." See aquage.

aqueduct (E & Sc.) From the L *aquaeductus*, 'to lead, to bring'. 'An elevated structure for conveying water, such as that in a canal or brook, over a railway, road or river', or even one drain over another, such as Cowbridge Aqueduct, in Lincs, which carries Stonebridge Drain over Cowbridge Drain. Leland (5.57) uses the form: "aquaduklyd" and Phillips (1720) gives: "a conduit or passage for conveying water from one place to another." The most famous aqueducts are, of course, the ones built by the Romans for water supply, in particular, the famous Segovia aqueduct in Spain, with a length of about 15 km, used to supply the city with water. Very few aqueducts are shown on the modern map in Britain, one of the best is in Scotland, shown between the 160-170 metre contour line on OSX 341 at NS2373 and a Roman aqueduct can be traced, north of Hadrians Wall at NGR NY7267. See water bridge.

aqueductile (L) See aquaductile.

aqueductum (L) See aquaductum.

aqueductus (L) See aquaeductus.

arc (Sc.) The DSL (SND) state that the term arc and ark is sometimes used for: "The whole of the waterway from the end of the mill-lade to the tail-race."

arched-drain (E) An agricultural drainage term listed by Ogilvie (2.96a). Rees (Vol. 34 under Sub-Soil) records it: "arched main-drain."

arched main-drain (E) See arched-drain.

ark (Sc.) see arc.

arm (E) 'A branch of a river or stream following the division of a river into smaller channels'. At least two 'arms' can be seen on maps – at NGR TF2909 and TQ9730. Leland frequently uses arm in his *Itinerary*, in the form "arme" – "The ryver of Severne brekethe into 2. armes in the medowes a litle above Glocestar" (2.63). See armlet.

arme (E) See arm.

armelet (E) See armlet.

armelette (E) See armlet.

armlet (E) The dim. of arm, an armlet is basically 'a small channel of a river or stream following the division of a river into smaller channels'. Leland (2.89) uses the form "armelets" in his section on the River Stour in Worcs: "Here dothe Stoure ryver breke into 2. or 3. armelets, and servythe milles," and "armelettes" appears a number of times elsewhere (1.111, 137 and 274). See arm.

arrey (Mx.) Kelly (1866:11b) defines this rare term as: "a millrace."

asc (Brit. & OGael.) See ascaig.

ascaig (SGael.) Mackay (189), under Loch-ascaig, states: "ascaig; escaig, dim. of asc; esc, little stream or small brook, lake of the small stream; asc, esc, esk, ask, are British and Old Gaelic terms." This would appear to be the only mention of this term, further, under Loch Traderscaig (190), he gives: "O.G., truid, stripe, battle, *air* ou, and, *scaig* contraction for escaig, dim. of esc., small stream." The esc term is also listed in the (CL:EOI). See wysg.

ask (Brit. & OGael.) See ascaig.

assuera (L) CTM (191a) gives: "drain", the connection with sewer can easily be seen here, and (190b) "asseware, assewiare, to drain marsh ground."

ath (SGael?) Only in Milne (1912:1) as: "stream, ford." As with so many of Milne's terms, no other authority seems to support the fact that 'ath' exists as a generic term.

au (Egyp., G, Ir., SGael? & Wel.) Joyce (1869:3:54) states: "au, aw, ow, either separately or in combination are the names of rivers all through Ireland, representing in sound the original Irish word *abh* or *abha*." Spenser's *The Faerie Queene* (235a, Canto X1, xli, line 2), in one of the verses describing Irish rivers, gives: "the stony Aubrion". Although Jamieson (supp. 1a) gives au as one of the forms of abh, water, no particular instance of its use is given. Waddell (1927:2a) representing the

Wel., also lists au as: "river, sea" and Budge, in his *Egyptian Hieroglyphic Dictionary*, (1:31b) lists it too – defined as: "river, stream." In Gmy., Au(e) is still used of a stream or channel and in Cambodia, au is used as a common prefix for stream terms such as Au Avan. See USBG (GFR:iiia); aw and ow.

auan (Co.) Lhyud mentions this form no less than five times in his *Archaeologia Britannica* – Pryce just once, as a torrent or land flood. See avon.

aub (OIr.) Another variant of the ab stem given by Hessens (177b). See oub.

auld (SGael.) See allt.

ault (Mx.) See allt.

aun(e) (Bret. & Co?) For the Bret., Lhyud (22c) states that *aun* is Armorican (Bret.). For the Co., Jago (1887:133a) gives: "aun" from Williams but Williams actually gives: "auon" (1865:15b). In view of this, aun, as a generic, must be considered a Bret. variant. The RN forms for the Dev. Avon, given by Leland (1:217), are "aune" and "awne". Rowe (14) gives: "The Avon, Aven, or Aune (which seems to have been the antient appellation)" and Crossing (1:64), particularly states that the Avon: "is always called the Aune on the moor". See auon, auney, avon, and awn.

auney (Co.) Dexter (1926:37.93) lists two words under the PN Inney which he defines as: "Little river (Inney, Auney; dim. of Aune and Avon)". The, apparently, singular example of these words by Dexter does not appear to be confirmed by later authorities and are, therefore, questionable. The term is identical to that of Taylor (197) but, again, unconfirmed later. Weatherhill (34b) questions the PN Downinney, which might belong here, but dowrhyns and dowrhens might have a claim too. There is also a R Inney (Dowr Enni) in Corn., a tributary of the R Tamar. See aun(e) and Padel (99).

auon (MBret., OCo. & Wel.) The MBret. form is given by George (2009:80b), under avon. Holder's MBret form (1:9a) is: *auonn*. The Co. word comes from the Old Cornish Vocabulary – *Vocabularium Cornicum*, a 12th-century glossary which can be found in Norris (311-432) and the 1853 Zeuss edition which gives, (2:1119): "fluuius, auon." Drayton uses auon, on his map of Hampshire and Borlase, Williams (1865) and Jago all list it too – Ferguson (24 & 26) and Taylor (196) actually spell it: "auwon"! See afon; avon and, for the early use of the auon form in Wel. texts, especially the entries in the *Red Book of Hergest* (c14th C), Skene (1868:2:303 line 11 & 2:384). In addition, the LL (359) has "auon" as a variant given in the British Museum Ms. Cotton Vesp. A. Xiv.

auonn (MBret.) See auon.

auwon (Co.) See auon.

avan (Ir. & Wel?) Lhyud (42:2c & 290a), in both instances, gives this form as a derivative of the L "amnis", Wel., "avon" and Co. "auan". Glamorgan (6:2217) mentions the River Avan and: "myll pounde of Avan" as well as other Avan mentions elsewhere. The closest we can get to a tautological *'river-river'* in Wel., is, therefore, probably the River Avan, (Wel., Afon Afan) which flows into the sea at Port Talbot, shown at NGR SS8496.

avana (Skt.) See abhainn.

avani (Skt.) See abhainn.

aven (Bret.) Used by some writers as a variant form of the RN Avon. See aun(e) and avon.

avin (Celt.) See abhin.

avon (Bret., Co., E?, AIr. Sc. & Wel.) All the avon terms, RN's and associated PN's go back to the OBrit. *Abona* and *Abone* (Abona, Ravenna, RN; Abone, AI, PN) – PC **abon-*, and, ultimately, the PIE root **ab-*, **ap-*, a stem which can be traced to the Sanskrit *apas*, 'water'. According to Matasovic, the Gaul. ambe, 'river, rivo or rivos' belongs here too; compare Skt., *ambu*, 'water', Williams (1872:78c). Cognates are: Bret. *aven*; Co. *avon*; MBret. and OCo. *auon*; OIr. *ab*, *aub*, *ob* and *oub*; Ir. *abha*; SGael. *abh*; ModIr. and SGael. *abhainn* – AIr. *avon*; Mx., *awin*; L *amnis*; Lith. *upe*; Lett. *upe*; OPr. *ape* (recorded in the OPr.EV) and Wel. *afon*. In addition, the **ap-* root is found in Su., an extinct language closely related to OPr., which has *ape* and *apis* as terms for 'river', and, in the ancient Anatolian languages, we find *hapa* (Hittite) and *hapi* (Luwian) as terms for river. Another interesting fact is that Hapi was the name of the ancient Egyptian God of the Nile. Overall, the **ab-*, **ap-*, root is colossal; it extends well beyond Europe, through the Middle East, with its Avestan (Iranian) connections, to the shores of China and its ancient Tocharian Language; the B, 'Western Dialect' list gives: "**ap-*". So, this is one of the really big ones, probably the biggest of all. Over the last 2000 years, at least 120 'avon' synonyms and RN forms have appeared in print, from the time of Tacitus up to the present – all listed in the Appendix. The Bret., Co. and E avons; the Ir., avons and owens; the Mx. awins and owins; the Sc., avons and the Wel. avans, avons and afons are all synonymous; very often found side by side in documents and charters. For the Bret., Henry (21) gives: *aven* – "rivière" and Bret L (15) lists: *aven* and *avon*. In Bret., *rivière* and *ster* have replaced *aven* as a common noun for river. In Cornish, avon, as a generic, is not disputed in the dictionaries of Pryce, Williams or Jago, and Nance (1978:212a) brings the word into the 20th C. further updated by George (1998:12b & 2009:80b). In E, avon is not recognised as a watercourse generic but it should be. As Bradley (10), and many others before him, have pointed out: "It is clear, however, that Avon (in Old English Afene) was not originally a proper name at all. It is the British word for river ...." Most etymologists and toponymists now agree that it means nothing more than 'water, river', which has not always been the case. The general view is that early Old English speakers, on enquiring as to the term of a particular watercourse, may have been given the answer: 'avon', which they took as the proper name, instead of the generic. Arguments for and against this theory have existed for years, see ERN (1928:23); Lockwood (1972:174); Rivet & Smith (1979:238b-241b); Nicolaisen (2001:229) and James (āß). The E River Avons – six of them – are all discussed, together with lists of variant forms, by Ekwall in ERN (20-2). The OGS (1:93a-95a) mentions four Avons in

Scotland, two of which are discussed by CPNS (430). The contracted form, Abon, is printed in KCD (29) and in Skinner's *Etymologicon Linguæ Anglicanæ*, one of the earliest, if not, *the* earliest etymological dictionary, defined as: *"Britanniam fluminum nomen"*. The rare abbone form, is listed in BM (1:31) and the Wel. form of abona – *Abbona* – is attested in the LL (326 & 331). We would like to think that Tacitus (55-117 AD) in his *Annals* (12:31), was the first to actually record the name Avon, nearly 2000 years ago. The Latin Library (www.thelatinlibrary.com) and others, give: *"Avonam"*, which is supposed to be the form used by Tacitus. L & S, in *A Latin Dictionary*, differ somewhat and give: "Auvona, a river in Britain". We now know that, over time, these *Annals* have received various emendations to the Mss., thus rendering the forms questionable, so, although we would like to think that Tacitus first mentioned the name, for now, we can only accept the experts' view that this may not be the case. However, the fact remains that although The Latin Library still promote the *Annals* with the *Avonam* form, the Mss. are unreliable and cannot, therefore, be taken seriously, hence the reason, no toponymist, as yet, has ever credited Tacitus with the first mention of the name Avon. In Ireland, avon is used as a prefix or suffix in some RN's; Avonmore (otherwise Owenmore) for example – Chalmers (19) mentions that: "The Aufona river in Ireland, which is incorrectly written Ausona, in some maps, is obviously the Celtic Avon, the name of so many rivers, in Britain, which is merely Latinized into Aufona." Other avon RN's in Ireland are listed by Joyce (3:403): "Kilfahavon in Monaghan"; Hogan (2a & b and 458) and MIPN. See auon; ab-; am-; au-; av-; aw-; ob-; ot-; ou- and ow- cognate forms; Glamorgan (6: 6a, b, & 7a) for a sizeable list of Wel. avon forms prior to becoming afon and especially (6:2180) for Blaen Avan; the PNGl (4:98); Skene (1867:lxxxi, for the fantastic 'haefe' form); Watts (28a); Williams (1872:48b) and, especially, the *Nostratic Dictionary* by Aharon Dolgopolsky (94) for a full and very comprehensive proto list. Older references can be found in Pokorny (1:1), under ab-, and, for the controversial *Annals Mss.*; C. O Brink, *A Sixteenth-Century Editor of the Annals of Tacitus*, pp. 120-2, in The Classical Review, Vol. 64, No 3/4, Cambridge University Press, 1950.

avonig (Wel.) See afonig.

avon maga (Co.) See avon vaga.

avon vaga (Co.) George (2009:80b) gives: "avon vaga, tributary" and Nance (1978:180b) lists the variant "avon maga" under tributary. See ragavon and rhagafon, both tributary terms.

aw (Co. Ir., SGael. & Wel.) Waddell, (1927:2a, under A, Ā) lists 'aw' as Co., "river." For the Irish, Joyce (1869:3:54) states: "au, aw, ow, either separately or in combination are the names of rivers all through Ireland, representing in sound the original Irish word *abh* or *abha.*" Further references can be found in Bartholomew (34a) and Blackie (2). The latter roots this form back to Persian. For the SGael., Jamieson (supp. 1a) states: "The terminations au, aw, o, ow, are forms of Gael. abh, water; as in the Awe in Scot., and the Ow in Ireland. The Wel. references to aw can be found in Pughe and Pryse (1866:643a) and Morgan (86) discusses the term in defining the PN Hawen. Another interesting fact about aw is that it is used as a very common prefix for river names in Iraq. See au; awy; ea and ow.

awan (Co.) Williams (1865:16a) defines this variant as: "a river, torrent, landflood." and further states that Lhuyd's instance "Ternewan an awan" means a river bank. Lhuyd (3b) actually gives auan, not awan – "Terneuan an auan", However, the validity of the term is confirmed by Nance (1978:212a). See avon.

awedh (Co.) This term is listed by George (1998:13a & 2009:81b) and Nance (1978:212a) as "watercourse." Lhuyd (233b) gives: "awedhyr, running water". Compare Wel. aweddw(f)r, various dictionaries. The form aweth can be found in Bond, who also gives: "watercourse."

awell (Wel.) Evans (1852:54a) and Pughe & Pryse (1:242b) both give: "aqueduct", Spurrell's (34b) gives: "conduit", as does the GPC.

aweth (Co.) See awedh.

awin (Ir. & Mx.) Seward (1797), in his *Topographia Hibernica*, lists most of his RN prefixes using this term, as does Bartholomew (1904:34a): "Awinbeg, Awinea, Awinmore" etc., some of which now have alternative names – Awbeg for Awinbeg for instance. Hogan too, has three rivers listed using this term (1b, 2a & b). For the Mx., Kelly (1866:13a) gives: "a river." Only a few Manx watercourse terms have made it through on to the modern map, river has replaced most of them, awin is, therefore, one of the few and can be seen on IOMS at SC2473 and 3482. There are many references to awin in Moore's *The Surnames & Place-Names of the Isle of Man*, especially p 169; throughout Kneen and in Broderick (5). See owin.

awn (Co.) Harvey (109) lists awn as a "sedgy river" and Leland (1:217-8) uses the form awne, three times for the Dev. Avon. See aun(e).

awne (E) See aun(e).

awni (Ir.) A variant of avon, a ME form, probably by metathesis, for awin, found in Spenser's *The Faerie Queen*, (235a Canto X1, xli, line 5). Joyce (1911:80) identifies the 'Swift Awniduff' as the 'Blackwater, flowing between the counties of Armagh and Derry'. See abh, au, aw and oure.

awon (Co. & Wel.) The Co. term is listed by Bock & Bruch. Morgan (10) states: "Afon, a river, comes probably from the Celtic awon, the moving water." A term also found, and confirmed, in the dictionaries of Evans (50a) and Pughe & Pryse (1:243a).

awy (Co? Wel.) Waddell (1927:2a, under A, Ā) lists "awy as: "Co. & Celt., river." Borlase gives awy as: "the old word for river." His bibliography names the source as Rowland's Welsh, perhaps, therefore, not necessarily a Cornish word.

ay (E) See ea.

B

baach (Lux.) See bache.
bach (E, F, G, Lux & ME.) See bache and beck.

bache (E) The ME form of the OE *bæce*, a word, according to B-T (60) "which seems to occur only in lists of boundaries in charters", BCS 233 and KCD 154 (S 126) for instance. 'A stream, or the valley through which a stream flows', is the definition given in most cases. Holland, in his Chester word book (14) gives: "bach, a fall, or a stream, as in Sandbach", Leigh (9) has the same. It's a very common term in Ches. – see PNCh (5:1.i:99-100) for a huge list of relative PN's. In Layamon's *Brut* – there are three different forms, the notes (3:447), referring to line 757, give: "At Clent in Cu-bache" – line 2596 gives "bæch" and 5644 has "bæche". The *Cu-Bache* comes from *The Life of St. Kenelm* (dated 1305) in Furnivall's *Lives of Saints* (54 line 244, 245 & 266) which gives: "coubache" and (55 line 289) "coubage". The Shrops. and Herefs. dialect form is batch; two watercourses bearing this appellative are shown on OSX 217 at SO4088 and 4189 and there is a Batch Brook shown at NGR SJ6280. Bannister (11, under bache) gives a number of batch PN's. For a F example, see *La Sauer Rivière* – its variant form is *Sauer-Bach* and Lux. uses *baach*. However, these forms may be closer to the G *bach* than to bache. In Gmy., *Bach* and it's dims. *Bäch(e)l; Bächle* and *Bächlein* all mean 'stream, ditch'. A point to be noted here is that Gm. generics are always capitalized, unlike ours. For a full digest of this very widespread element see beck; DEPN (21a); The EDD(1:108); EPNE (23-4); the MED (under bach); PNC (312); PNCh (all vols.); PNDb (4); PNSa (2:161); PNWo (312) and USBG (GFR:iiia).

Bäch(e)l (G) See bache.

Bächle (G) See bache.

Bächlein (G) See bache.

bäck (Swe.) See beck.

back brook (E) Used of a by-pass course to the main stream for some necessary reason, much the same as backwater is sometimes used as a flood alleviation channel. Back brook was often used as a common term for the channel of water which by-passed watermills; owners were obliged to ensure that sufficient water was available further downstream for use by other mill owners. The old course of the River Lugg, in Herefs., is shown at NGR SO5538 as a 'Back Brook'. See the other back- compounds.

back burn (NIr.) Shown on OSNIDS at H4854 but probably in error for Blackburn nearby. Here though, not in the same sense as some of the other back- compounds.

back drain (Sc.) On OSX 348 at NS7578, a back drain is shown running alongside the Forth and Clyde Canal in Scotland. It presumably acts as some sort of recycler but, being connected to so many other drains, makes it difficult to establish its true purpose.

back gutter (E) This compound is shown at NGR NY8607. When back is used as a prefix, in watercourse terminology, it usually denotes some kind of diversionary feature as it does in 'back brook' or 'back water', but here no such feature is present. It may be the only showing of the term to be found on the modern map. See other back- compounds.

backie (Shet.) Jakobsen (31a) mentions this word, which comes from a small glossary in the *Description of the Islands of Orkney and Zetland* by Sir Robert Sibbald, published in 1711, which gives (31): "Backie a small running water which gave rise to the surname of the people of that Name." See beck.

back river (E) Stone (169), in his book, *A review of the Corrected Agricultural Survey of Lincolnshire by Arthur Young* (published a year after Young's survey) states: "The same principle of drainage, by means of a back river, I conceive, ought to have been adopted on the *south side* of the Trent." However, there does not appear to be any mention of the term 'back river' in Young's original survey (1799) but at least one can be found on the modern map, lying south of Peterborough, in Cambs., at NGR TL2097. See other back- compounds.

back-stream (E) Often mentioned as a necessary requirement of watermills, 'a watercourse used to carry off surplus water'. Elworthy (1886:21) states: "The leat [water-course] and the back stream are as indispensable as the waterwheel itself." See the Back Stream shown, north of Taunton, at NGR ST2027, and other back- compounds.

backwater (E, ME & Sc.) Heslop (27) has this listed as: "BACKWATTER ... the overflow from a mill race." Here, the dialectal 'watter' is used for water. The ME form 'bakwatere' appears in *Polychronicon* (1.57) by Ranulf Higden and two 'back waters', acting as by-rivers, can be found on maps – one at NGR SU5096, near Abingdon and the other at SU7880, near Henley-on-Thames. A Sc. backwater can be seen at NGR NO2558 and the term is often compounded with gang in Scot. See The DSL (SND); PNLei (3:48); PNRu (310) and other back- compounds.

back water gang (E) See backwater.

backwatter (E) See backwater.

bæce (OE) See bache.

bæch(e) (ME) See bache.

bæk (Dan.) See beck.

bage (ME) The Midlands form of bache, used as a suffix in FN's and PN's. Furnivall (55 line 289) gives: "coubage". Bannister (11) gives: "bage", from Bage Farm in Madley. See bache; PNDb (1:118) under Hazelbadge and PNWo (312) under Badge Court.

bah (OHG) See beck.

bak* (OLFrk.) See beck.

Bäke (G) See beck.

***baki(z)** (Gm.) See beck.

bakk(i) (SNn.) See beck.

bakwatere (ME) See backwater.

barrel arch (E) See culbit.

barrel-drain (E) An agricultural drainage term, one of the many -drain prefix variants listed by Ogilvie (2.96a). Loudon (505) discusses barrel-drains stating that: "bridges are frequently required on estates and farms, for crossing ditches and watercourses. They are generally large stone conduits or barrel-drains"; there is a sketch of one on p 707. Bedford (249:530) also mentions the term.

batch (E) See bache.
bec (OE) See beck.
becc (ME & OE) See beck.
bece (OE) See beck.
bech (Lux.) See beck.
becht (Lux.) See beck.
beck (E, G, SNn. & Sc.) From the ON *bekkr*, 'a brook or stream', very common in the eastern and northern counties of England settled by the Danes and Norwegians, and particularly well evidenced in the Norn language of Shetland. Grose (10) gives: "beck or beek" – a form found in Dut., Fle. and G (see below). In fact, I doubt if any northern or eastern word book omits the term – it's one of the most common generics used in those parts of the country. Parish (17) in his *Sussex Dialect* book gives "a rivulet", but PNSx (2:247) do not confirm this. The OE forms are "*bec, becc* and *bece*" – B-T (67 & 74), becc is also found in the ME *Cursor* Mundi – Morris (1874:2:515 line 8946). PP (29b) renders beck – "Bek watyr, rendylle, *riuulus*." WW (1:736.24) lists: "*Hec rivulus*, a bek." The word beck has, of course, many entries in all the classic volumes of state: *Book of Fees; Close Rolls; Subsidy Rolls* and so on – usually as personal name suffixes – see the MED under bek. ERN (28-9) discusses beck at length and PNC (312) has many forms. Being so widespread and for a full digest, reference to all PN books covering the appropriate counties is recommended. The SNn. examples, collected by Jakobsen (1928:1:31a) are "*bakk*"; "*bakki*"; "*bekk*" and "*bekki*" and Angus gives the Shet.: "*bek*". It's common throughout all the Gm. speaking countries: *bak** and *beke** is given as OLFrk; *beki?* and *biki?* as OSax. and *bah* as (OHG) by Köbler. The Gm. forms are **baki* and **bakiz* and the IG is **bʰog*. Den. uses *bæk*; *beek* is used in the Nth., and is also found in Fle., together with *beke*, a form given by Kiliani (48b). Gmy. use quite a few forms: *Bach; Bäke; Beck; Becke; Beek(e)* and *Bek(e)*. Swe. has *bäck*, and Switz., *bach*. The word has even crept in to Fra., in various forms: *bach; becque; beek* and *beeque*, and, in Lux. too, as *bach; bech; becht* and *bich*. For the rare

bich form, compare Wisbich, in Fenland Notes & Queries (1:322). So many forms – so many cognates – a good example of our relationship with the rest of the Gm. Languages. There are, of course, many becks shown on maps, particularly in the English Lake District and The Broads, and, OSX 161 at TQ3767, marks, what is probably the most southerly example. A Sc. beck is shown on OSX 336 at NT0430. See bache; backie; all the beck- stems; Macbain (1922:82) for the PN's Diebek and Dubec; The EDD (213a) for the dialect regions; EPNE (1:26) for associated PN's and the MED under beche, used as a surname.

Becke (G) See beck.

beckett (E) A dim. of beck, 'small brook'. Brockett (1.31) seems to be the only wordsmith to have listed this term in his *Glossary of North Country Words*, other than Dexter (1871:7b) who mentions the Co. PN Becket: "little (beck) brook."

beck-grain (E) According to Dickinson (20a) it is the place where: "a beck divides into two streams." See grain.

beckstead (E) See becksteead.

becksteead (E) There are slightly different forms given for this term – F. K. Robinson (1876a:14a) gives: "becksteead" and C. C. Robinson (1876b:44) gives: "beckstead", both mean 'the bed or channel of a brook'.

becque (F) See beck.

bedum (L) Martin (197b) gives: "the portion of a millstream which turns the wheel and is boarded up to increase the force of water."

beek (Dut., E, Fle., F, & G) See beck.

Beeke (G) See beck.

beeque (F) See beck.

bek (G, ME & Shet.) See beck.

beke (Fle. & G) See beck.

beke* (OLFrk.) See beck.

beki? (OSax.) See beck.

bekk(i) (SNn.) See beck.

bekkr (ON) See beck.

beum (SGael.) Armstrong (61b) gives, amongst other things: "a stream, a torrent". There is an Allt a' Bheum shown at NGR NH2493 but this, of course, carries its own qualifying term.

*bʰog (IG). See beck and brook.

bhran (SGael.) See bran.

bich (Lux.) See beck.

biki? (OSax.) See beck.

bior (Ir.) See bir(or).

biorchli (Ir.) O'Reilly (64a) defines this as: "a stream of water." Cormac (27, under Additional Articles) gives bircli "waterstream." See bir(or).

bir(or) (Ir., Mx.) This might be related to biorchli, possibly 'a stream', but see eDil's entry. Hogan (115b) gives bir and Cormac (19, under biror) gives: "bir a well or stream." For the Mx., Kelly (34a) gives: "byr a water, a river from ar, a river. Ir bior." The bir term is also listed in the (CL:EOI).

bircli (Ir.) See biorchli.

bog (E) From the Ir., *bogach*, 'a bog'. This term is questionable, but a bog is shown on a number of maps as a generic term at NGR's SU1709; SE0995 and NZ6613. Another bog is marked in black at SU3307; bog is not usually defined as a watercourse, but the fact that the OS have marked at least three of these courses in blue make it so.

bog-drain (E) This is 'a collective term for all the small ditches and drains used in bog draining', given by Stephens (1848:16).

boorne (E & Sc.) See bourn.

bord-ríþig (OE) B–T (101) give: "a stream running in a channel made of planks (?)."

born (E) See bourn.

borne (OPr.) See bourn.

botag (SGael.) Watson (1904:lxxxvi) states: "Botag is a wet or soft channel in a peat moss." It's mentioned again on (101) as "a sun-dried crack, or narrow channel" and (215) repeats the first definition. Dwelly (111a) gives: "botach, reedy fen, bog", and the nearest Irish mention is to be found in O'Reilly (72b) who gives: "bothach, s. m. a bog, a fen, a marshy place." The foregoing suggests that botag could be considered as a watercourse generic, especially as the word 'channel' is mentioned and it is obviously associated with wet places.

bothem (E) A variant spelling of bottom, Wright (1880:242a) defines it: "A watercourse." See bottom.

botm (OE) See bottom(e).

botme (ME) See bottom(e).

botom (E) See bottom(e).

bottom(e) (E) From the OE *botm*, the ME form is botme, given by Wycliffe (3:258b, from Isaiah 19:7): "The botme of watir ...." Leland (2:155) uses the form botom and Lambarde (239) mentions that the River Stour in Kent, together with other streams: "all passe in one bottome to Wie, and to Canterbury," and further (260), he gives: "Wels (or springs) the which creepe at the first out of the earth, and bee conveied in slender quilles, then afterwarde (meeting together in course) doe growe by little and little into bigger pipes, and at the last doe emptie themselves into some one bottome, and so make up a great streame, or chanell." Here, he seems to be using 'bottome' as a watercourse term, but, irrespective of this, bottom is not disputable because there are quite a few shown on maps, as definite watercourse generics – one on OSOL 22 at SZ1995 and another on OSX 158 at SU4067. Two more are shown on OSX 194 centred on TL4041. The grid designations are miles apart proving widespread use of this term.

bottom carrier (E) An agricultural drainage term given by FD – "bottom carrier or tail drain." A recent, more modernistic term perhaps, as it does not appear to be given by any of the other drainage authorities listed herein. See carrier.

bourn (E & Sc.) From the OE *burna* and *burn(e)* found mainly in the south of England. The 13th C, *Old Prussian Elbing Vocabulary* (OPr.EV) lists "*borne*" – "water spring", on folio 170a line 64. Ashton (88) has "boorne", the same as Lambarde (260). Heslop (85) gives: "born" as does Jamieson (Supp., 49b) for the Sc. There are a number of mapped instances – OSX 211 at TL9468 and OSX 145 at TQ0147 are just two examples.

See burn for the etymology and ERN (41-4) which lists many bourn stream names.

bout? (E) This term is normally used in ploughing, however, there is a watercourse shown on OSOL 19 at NY8720 bearing this appellative. Further references are lacking except for a Scottish definition given in The DSL (SND) under bowt.

box drain (E & Sc.) Ogilvie (318) defines this agricultural drainage term as: "An underground drain regularly built with upright sides and a flat stone or brick cover, so that the close section has the appearance of a square box." Loudon (707) gives: "the walled or box drain", and further, "The boxed and rubble drain." Further still (1247), he defines sough as "a box-drain." Scottish definitions are comparable and come from Jamieson (1.271a) and Warrack (49a). These terms are all interconnected and sometimes confusing depending on which authority gives them; see dribble drain for a series of alternative -drain terms.

br. (Ir.) See brook.

braga (Ir.) O'Reilly (73b) defines this term as: "a stream of water." Nennius (200.5) gives: "braigit" and (201a & b) states: "The word braigit denotes the sluice or narrow canal through which the water flows from the linn or pond upon the wheel of the mill. Mr. O'Donovan informs me that these words are still so used in the County Kilkenny, and probably in most other parts of Ireland."

braigit (Ir.) See braga.

bran (SGael.) Dwelly (114b) gives: "Mountain stream, name of several rivers in the Highlands." Milne has a number of bran variants: "bhran" (1912:141); "branan" (51); "braon" (19) and "braonan" (61). As mentioned above, some of Milne's generic terms are not verified by current scholarship, but here, we do appear to have a valid generic term. According to Watson (1904:165) bran: "is an obsolete word meaning raven. As applied to a river, the reference is not very clear, but it may have been given simply from ravens having haunted some parts of it."

branan (SGael.) See bran.

branch (E) Ogilvie (1:322b) defines branch as: "a river running into a larger one, or proceeding from it." Two 'branches' of the River Avon are shown on OSX 168 – the Sherston (Wilts.) branch at ST8386 and the Tetbury (Gloucs.) branch at ST8991.

branch-drain (E) An agricultural drainage term given by Burke (1841:12) and Stephens (1848:34).

branch float (E) An irrigation term given by Marshall (1796a:135-6). See float.

branch trench (E) An irrigation term given by Boswell (37) who states: "branch trenches, taken out of the other trenche." Loudon (729:4439) gives "... to force the water into the branch trench ...."

braon (SGael.) See bran.

braonan (SGael.) See bran.

brick-drain (E) An agricultural drainage term. Loudon (708) states: "The brick drain is formed in a great variety of ways, either with common bricks and bats in imitation of the boxed and rubble, or rubble drain; or with bricks made on purpose, of which there is great variety." Stephens (1848:138-9) has a paragraph on brick drains too.

brigstone (E) Atkinson (1891:4) states: "A brigstone is a kind of rough, conduit for water across a gate-stead".

brim (E, ME & Sc.) From the OE *brim*, 'sea'; ME *brim*, 'river, stream'. Wright (1880:257a) gives: "a river." B-T (126) give the OE compounds: "brim-streám, brym-streám, *a stream, river*". The ME form given by Madden (371b) and Morris (1869, line 365), is *brymme*. Sc. definitions are to be found in Jamieson (Supp., 58a & b) and The DSL (DOST), under brim and brym. The references therein come mainly from Barbour and Henryson. Barbour (256 line 339) gives: "Lawch by a brym, he gert thame ta" and Henryson (217 line 9) gives: "For Goddis lufe, sum bodie ower this brym" and (218 line 38) "And thocht the brym be perrillous to waid". There is a Brim Brook on Dartmoor, Dev., shown at NGR SX5887, possibly the only instance of brim, in any form, on the modern map. See the

MED and PND (3).

brim-streám (OE) See brim.

brin (Shet.) See bryn.

bróc (OE) See brook.

broek (Fle. & Dut.) See brook.

brok(e) (E, G, LG, ME, OSax & OSc.) See broken, brook and rinnon.

broka(m) (Gm.) See brook.

broken (ME) No doubt a dim. of broke; Halliwell (213a) mentions the term: "BROKEN. A brook. Skinner." A rare term, also found in Godstow (554, line 27). The nearest Wright has (1880:259a & b) is: "broke and broket".

broket (E) See broken and brooket.

brokwater (ME) PP (50) list "brokwater" with the corresponding L terms, "riuulus, torrens".

brone (E) Pulman (1875:598) gives this rarity, as a variant of burn, in the following statement: "but in military affairs the term Brunenburgh is invariably given to it, a term which designates at once the river and the fortress upon it – burn, or brone, a brook or stream, and bury, a place of retreat or defence". A form, probably by metathesis, for borne.

brook (E, Ir., NIr., OSc. & Wel.) From the OE *bróc*, B-T (106 & 126). ME forms are brok(e); bruke; bruche and brooke. The most common term applied to minor and some major watercourses in England, however, in the Gm. langs, brook means 'water meadow, marsh, pasture' and was brought here and only later applied as a generic term in A-S days. In Kent, in the 19th C, it was still being used of a water meadow! See Parish & Shaw (1888:20) under brooks. It is still used in its original meaning in the Gm. languages to the present day, but never as running water in general. The corresponding terms are: LG *brok*; OHG *Bruoh*; MHG *Bruoch*; G *Bruch*; Fle. *broek* and OSax., *brok*. The Gm. terms are *broka* and *brokam*, and the IG is *bhog*, (Köbler). In Gmy., *Brok*, *Bruch* and even *Brook* are still in use, but only for 'pasture or marsh'. On OS maps, brook is common from the very north of England to the tip of

Cornwall, and is also found in Ire. Scot. and Wal. In Ire., on OSIDS 56, brook is extremely common, often contracted to br. We even have a Brown's Beck Brook showing on OSIDS 56 at S9498. In NI, brook is shown on OSNIDS 4 at C8034 and OSNIDS 29. For the Sc., Jamieson (58a) gives: "bruk, bruke, s. A brook, stream", and bruke and brok(e) can be found in OSc., probably due to English incursions, but not in ModSc. In Wal., particularly in South Pembrokeshire, brook appears more often than the most common Wel. term nant! One possibility is that this may be due to Fle. influence – *broek* is still used in Bel. today, but, whatever the answer is, there has to be some sort of explanation for its frequency of use because brook is not found in Wel. dictionaries. There are other instances of brook in Wal., one is shown in the suburbs of Cardiff on OSX 151 at ST1978 and another near Borth on OSOL 23 at SN6191. Virtually all EPNS books list PN's with a brook element and there are entries in Sc. and Wel. PN books and elsewhere too. See The DSL (DOST); Nostratic (322:253); The OED1 and USBG (GFR:iiib).

brooke (ME) See brook.

brookelet (ME) See brooklet.

brooket (E) Camden (2:207) gives: "brooket"; Lelands form (1:301) is: "broket". The term is an earlier form of brooklet.

brooklet (E) Holinshed (1:120) gives: "brookelet" a ME form. Cornish (177) mentions: "Usually these brooklet valleys are choked with brambles or fern, and filled with rank undergrowth" – a later form of brooket.

brosnach (Ir. & SGael.) Dwelly (129a) and Lhuyd (329c) both give this term defined as: "river." Meyer (270) states that it is: "*a river-name*", probably referring to the one marked on OSIDS 48 at N2232.

Bruch (G) See brook.

bruche (ME) See brook.

bruk (Sc.) See brook.

bruke (ME, OSc. & Sc.) See brook.

brunna (Got.) See burn.

brunno (OHG, OLFrk. & OSax.) See burn.
brunnr (ON) See burn.
Bruoh (OHG) See brook.
brushwood drain (E) An agricultural drainage term mentioned by Loudon (4289-90).
brym (Sc.) See brim.
brymme (ME) See brim.
brym-streám (OE) See brim.
bryn (E, Shet.) Middendorff (20) gives: "bryn". Wright (1880:262) lists: "Brynnys, bourns; streams" and Edmondston (12) gives the S & O form "brin, a brook or rivulet."
bual (Ir.) Meyer (283) gives: "a stream, water." The term is also listed in the (CL:EOI).
buinne (Ir. & SGael.) O'Reilly (87b) defines this term: "a stream, a rapid river." According to eDIL it has more to do with flowing water rather than as a generic term. For the SGael., Dwelly (140a) gives: "stream." The term is also listed in the (CL:EOI).
bunny (E) There are many references to bunny – Wright (1880:270a) gives: "a sort of drain"; Long (8) defines it: "A small covered drain, or culvert, generally in front of a gate at the entrance to a field." Parish (23) adds: "... also called a cocker" and Cope (12-3) states: "The chink or narrow rift in the cliff-line, called in the Isle of Wight a chine, is known in the New Forest as a bunny." Dartnell (19) only defines it as: "a brick arch or wooden bridge". See cocker.
burna(n) (OE) See burn.
burn(e) (E, Mx., NIr., OE & Sc.) As with the E bourn, burn goes back to OE *burna*, cognate with OFris. *burna*; OHG, OLFrk. and OSax. *brunno*; ON *brunnr* and Got. *brunna*, all from Köbler. B-T (136) give: "burn, burna, burne and (140) byrne." The term is commonly found in northern England and some regions of Scotland, with many mapped instances, but it is also found in the English Midlands on Cannock Chase and next to the R Blithe in Staffs., shown on OSX 244 at SJ 9718 and OSX 244 at SK1117 respectively. There is another burn on

Dartmoor shown on OSOL 28 at SX7164. Surprisingly, at least
one example of OE *burna* survives on the map today – OSX
156 at ST9776 shows Cade Burna, a name which was first
attested in an A-S charter – BCS 717, (S 1575) – over a
thousand years ago. This is the oldest known instance of any
OE watercourse generic extant on the modern map! There is
another burna, Hreod Burna, not shown on the OS Map;
however, it is mentioned in an A-S charter of A D 1008:
"*andlang Hreódburnan on Worf*" in KCD 1305 (S 918) and Earle
(1888:392), this too, has also retained its original Anglo-Saxon
name to this day. The IOMS map shows two burns at SC2672
and SC3175 and in NI, burn is very common indeed – well
over 40 can be seen on OSNIDS 12; 10 on map 4 and many
more on map 13 as well as some of the other OSNIDS maps.
For the Sc., Jamieson (1:338b) gives: "a rivulet, a brook", and
in Wyntoun's *Cronykil*, line 4590, can be found the rare "*bwrne*"
form. BCS 34 (S1165) is probably the earliest charter to give
burne, dated 675 or before and the earliest *burne* given by WW
(1:29.27) is from an *Anglo-Saxon Vocabulary of the 8th Century*.
OE *burna* features in the majority of the EPNS PN Vols., and
ON *brunnr* is covered in PNNt (262) and PNYW (7:163). See
bourn; B-T, under burna and burn; ERN (41-4); all the Köbler
dictionaries and Pokorny.

burngate (Sc.) Only in The DSL (SND), under burn: "*burngate*, a
small water-course".

burn-grain (Sc.) This compound is given by Jamieson (1:339a)
defined as: "a small rill running into a larger stream." See grain
and the PN Burngrains shown at NGR NJ6555 where two
streams meet.

burnie (Sc.) Jamieson (1:339a) gives "burnie" and "burny" as
dim. forms of burn. A mapped instance can be seen on OSX
396 at NO7279.

burny (Sc.) See burnie.

burster (E) Gower (1876:84) defines this as: "a drain under a
road to carry off water", which is given in a 1641 Court Roll
document as "burstow."

burstow (E) See burster.

bush-drain (E) An agricultural drainage term. Defoe (2:174-5) mentions this term practised in Hertfordshire: "... by draining off the rain-water, which stagnated on the clayey surface, as in a cup, and chilled the roots of the corn; an invention, called bush-draining...." Marshall (1873a:46b) also mentions the term: "Bush-draining, underdraning (being done with bushes)."

bwrne (MSc.) See burn.

by- (E) Used only as a prefix in compounds denoting 'a channel cut to convey surplus water', usually one running close to a river or stream, but deviating from the main course for some specific purpose such as flood alleviation, overflow or as a diversionary channel to feed a watermill, generally 'a lesser stream'. The by-, in Byfleet, Surrey, is 'place by the fleet' and is not, therefore, used in the same sense as other by- compounds. See the various by- and back- entries and the OED1 (1:1232c & 1233a).

by brook (E) No references have been found confirming this as a generic by- stem term which appears only as a proper name shown at NGR ST8475. It joins the Bristol Avon near Batheaston. See ERN (443) and PNW (4).

by carrier (E) An irrigation term given by FD and also by Wikipedia, under Water-meadow. See carrier.

by-channel (E) Three by- terms are given by Bates in his book, *The Naturalist on The Amazons*, describing a number of watercourses; by-channel (496-7); by-stream (178, 503 & 506) and by-water (159 & 260).

by-dyke (E) Addy (35) gives: "a feeder or narrow stream for a mill-dam."

bye-wash (E) Ogilvie (361c) gives: "A channel cut to convey the surplus water from a reservoir or aqueduct, and prevent overflow." A by wash is shown on OSOL 31 at NY9619 and a bye wash on OSX 299 at SE5477.

byfleet (E) See by-.

by-lead (E) Ogilvie (361c) gives: "A channel cut to convey the surplus water from a reservoir or aqueduct, and prevent overflow."

byr (Mx.) See bir.

by-riuers (ME) See by-river.

by-river (E) Holinshed (1:181) states: "if you go backeward, contrarie to the course of my description, you shall find it so exact, as beside a verie few by-riuers ...." The same term is repeated in Harrison (1877:vi).

byrne (OE) See burn.

byset (E) Easther (20) defines this term as: "a channel cut in the road to take off the water."

by-stream (E) See by-channel.

by wash (E) See bye-wash.

by-water (E) See by-channel.

C

cabar (SGael?) Only in Milne (1912:240) as: "branch of a burn." As with so many of Milne's terms, no other authority seems to support the fact that 'cabar' exists as a generic term.

cael (Ir.) Joyce (1869:3:395, 396, 479 and 551) lists a number of PN's to which this term applies. O'Reilly (591b) states: "a stream flowing through a marsh." Hogan (135b) gives the same.

cafn (Wel.) Pughe & Pryse (1:384b) list this term as: "a passage for water, a channel", it also has a number of compounds, such as cafn dwfr, "channel or conduit"; -gwyllt and -melin, both "mill race" etc., all listed in the GPC (387b & c).

cafn dwfr (Wel.) See cafn.

cafn gwyllt (Wel.) See cafn.

cafn melin (Wel.) See cafn.

cahenryd (Co.) See chahen rit.

cais(e) (Ir. & SGael.) O'Reilly (96a) defines cais as: "a torrent, a stream." Dineen (1900:30) gives caise, which he defines (31) as: "rivulet." For the SGael., Dwelly (153a) gives: "caise, stream of water." See Hogan (142a) and (167a & b) casse. The term is also listed in the (CL:EOI).

camlas (Wel.) The GPC (402c), Spurrell's (67a) and Walters (1:555a) all give variable watercourse definitions.

canal (E, NIr. & Sc.) From the F *canal*, L *canalis*, 'a channel'. In ModL, canalis can mean 'a pipe or conduit' as well as 'a canal'. One of the main definitions (and probably the most well known) is 'a man-made channel built for the purpose of inland navigation, or to unite rivers and seas', such as the world famous Panama and Suez canals. The term does not readily present itself, or come to mind, as a watercourse generic in its own right, yet, a number of canals do exist as such in England, NI and Scotland. At least five 'canals' have made it on to the modern map: OSX 285 at SD3312 and OSX 141 at ST4551, give two in England, the latter, surrounded by a number of 'rhynes' on the Somerset Levels near Cheddar. In County Londonderry NI, a canal is shown on OSNIDS 14, at J9797, a short stream about 2 km. in length, which rises near Bellaghy and falls into Lough Beg. The southern Irish form is chanáil, shown on OSIDS 50 at O0337. For Scotland, three such courses exist, on OSX 311 at NX3950, OSX 361 at NR9030 and OSX 438 at NH8074, all as generic terms. The canal term is also used of 'an ornamental piece of water', usually quite narrow and linear, such as the one that existed in St James's Park, London, prior to its conversion into the more attractive lake we see today. A 'canal' still exists in the grounds of the Chillington Hall Estate, in Staffs, lying next to 'The Pool'. This too, is a narrow, linear water feature, viewable on OSX 219 at SJ8606.

canalibus (L) This term is the plural of canalis. WW (1:11.2) give; "waeterðrum", from an *Anglo-Saxon Vocabulary of the 8th Century*, the oldest document listed. Lindsay (1921a, line 3) gives: "waeterdruum". See canalis.

canaliculus (L) See canalis.

canalis (L) WW list a variety of OE meanings from the various vocabularies; (1:147.34): "mylentroh and þeote"; (191.5): "colimbus or aquaeductus, wæterþeote"; and, (198.25): "þruh uel mylentroh". So, we have 'aqueduct; water pipe; conduit;

mill trough or stream and channel', mostly from B-T who use the WW vocabularies for the majority of their definitions, thus leaving the final choice open. L & S (276a) give the dim. *canaliculus*. See canalibus.

canel (A-N & ME) This is not too strong a contender – obsoleted by channel and, further, was commonly used more as 'a street gutter', surviving in this sense as kennel. The same development took place in Wel. Definitions, as a watercourse, do exist – Riley (1862:3.300b) gives: "canele. F. A channel, or kennel, of a street, a watercourse ... English a canel", here though, not necessarily a natural watercourse. His L form canellus (380a) translates "kennel." Wright (283b) gives "channel", and Moisy (155), for the A-N, gives: "canal." Further definitions can be found in PN books and, for a complete overview, see PNCh (4:173) and (5:1: i and 128); CPNE (37) and PNYW (7.167).

canele (F) See canel.

canellus (L) See canel.

canol (Wel.) This is listed in the GPC (414c), defined as: "channel, stream, canal." It's The Wel. equivalent of the E, 'canal'.

caoch (SGael.) Milne (1912:49) states: "There is in Gaelic a word caoch, blind, and there had once been another meaning burn, with its dim. caochan, small burn." He gives Blind Burn in Aberdeenshire as a PN example. The term is rightly associated with blindness, so 'blind' or 'hidden stream' would be a suitable definition.

caocha(i)n (SGael.) Milne gives this as a dim. of caoch. Dwelly (161b) gives: caochan; caochain and caochanan all defined as "streamlet", and (162a), caochlain and caochlan as: "swift rill or rivulet." King (80) has a very comprehensive analysis of the term caochan and a mapped example can be seen on OSX 420 at NJ3428. The only other dim., shown on OSX maps, is caochanan, which can be found on OSX 394 at NN9488.

caochanan (SGael.) See caochan.

caochla(i)n (SGael.) See caochan.

caractis (L) This term is the plural of cataracta 'waterfall', so too is cataractes. WW (1:200.14) give: "*caractis*" and OE "wæterþruh", from an *Anglo-Saxon Glossary of the 10th Century*. Lindsay (1921b:32) gives: "ca(ta)ractis" and (1921a:110), "cataractes" and, the OE "uaeterthruch", which suggests E 'pipe, conduit'. See wæterþruh.

cardyke (E) See fen-dyke.

carf (NIr. & Sc?) This is listed in The DSL (SND) but Patterson (16) gives it as a NIr. term: "... a ditch; a shallow channel cut in peat bogs for conveying water." See cash.

carog (Wel.) See carrog.

carriage (E) An irrigation term given by Britton (32a) who states: "Never used but with the addition of *water*, as a *water-carriage* (Grose)." Cope (14) and Dartnell (23) both mention it as a drainage term and Loudon (4408) gives: "A carriage is a sort of small wooden or brick aqueduct, built open, for the purpose of carrying one stream over another, and is the most expensive conveyance belonging to the business of watering." So, for this term, definitions differ somewhat, especially that of Loudon.

carriage-gutter (E) An agricultural drainage term given by Elworthy (1886:116): "The main drain into which the branches in draining a field are made to run." See carriage.

carrier (E) An agricultural drainage term. Dartnell (23) gives: "Carrier, Water-carrier. A large watercourse." Dickson (1805-7:2:433) mentions the term a number of times and one is shown on OSX 247 at SK8231.

carrier ditch (E) An agricultural drainage term given by Young (1797:157-8) in the following statement: "I strongly recommend these carrier ditches to be open, though at the expense of a whelm at the bottom of a field where a cart-way is necessary; the leading many drains into one carrier ditch, that is covered, must be more liable to accidents and injuries, than where every drain empties itself singly into an open ditch."

carrog (Co. & Wel.) Defined by Pryce as: "a brook", it's the same in Wel., Evans (1852:191a) uses *carog* and Lhuyd uses the k- form "*karrog*". The GPC has "*carrog*". The Wel. RN's, Carog and Carrog, can be seen on OSX 215 at SN8097 and OSX 254 at SH4857 respectively, both prefixed *afon*, which suggests tautology and Carrog as a PN can be seen at NGR SJ1143. See karrog; karrag and DPNW (74) for Carrog and (169) for Glyncorrwg.

cash (NIr. & Sc?) This is listed in The DSL (SND) but Patterson (17) gives it as a NIr. term: "a covered drain made to leave a passage for water in wet ground or bog." See carf.

cast (E & Sc.) Only found in the very north of England and southern Scotland. Heslop (137) defines it as: "a mound of earth cast up as a boundary of lands between different proprietors, or as a fence. It also means a long ditch." No mention here of a watercourse generic but, Jamieson's (Supp., 5.69b), makes a definite connection: "A trench, ditch, cutting, or other channel for the passage of water." A number of maps show cast as a watercourse; English on OSX 332 at NU2319 and 340 at NU0742 and Scottish on OSX 367 at NT1383. For further Scottish definitions see The DSL (DOST & SND) under cast.

catadurpa (L) WW (1:800.24) give: "*Hec catadurpa*, cundythe", from a *Pictorial Vocabulary of the 15th Century*. This L term is usually defined as 'waterfall' and although it is given as corresponding with the E conduit here, 'cataract' would be better. There is, on the R Nile in Egypt, a waterfall called Catadupa. Cicero (106-46 BCE) mentions it in his *De Republica* (6:19).

cataractes (L) See caractis.

catch-drain (E) Ogilvie (1:414a) sums up this term comprehensively: "A drain along the side of a canal or other conduit to catch the surplus water. 2. A drain running along sloping ground to catch and convey the water flowing over the surface. When a meadow is pretty long, and has a quick descent, the water is often stopped at different distances by catch-drains so as to spread it over the adjoining surface." See all the catch- terms.

catchwater (E) Cole (26) states: "A drain cut to catch the water from higher ground, and carry it into a main drain without flowing over the lower lands." One can be seen at NGR NZ0247. See all the catch- terms.

catch-water course (E) An alternative name for a catch water drain. CEAJ (257b) gives: "I shall now proceed to describe the mode of discharging, by catch-water courses or drains ...." See all the catch- terms.

catch-water drain (E) Young (1799:226) and Knight (503a) both mention catch-water drains which can be seen at NGR SE0810 and TL5968. See all the catch- terms.

catch-work (E) Ogilvie (1:414b) gives: "An artificial water-course or series of water-courses for throwing water on such lands as lie on the declivity of hills; a catch-drain." See all the catch- terms.

cat-strand (Sc.) Warrack (77a) gives: "a very small stream."

cerrynt (Wel.) Spurrell's (77a) gives: "current, watercourse", GPC (468c), "current" only. Lhuyd (60b) gives two compounds – "hyttynt dur" and "kerrynt dur" as: "a river, stream or watercourse."

ceuffos (Wel.?) Pughe & Pryse (1:384b) give: "a gutter, a passage for water, a channel" and (329b) "a hollow ditch". The GPC definition is closer to a sink or gutter, therefore, not necessarily a natural or artificial channel used for fresh water, moreover, a term, perhaps, to be associated with sewers. See ceunant and ffos.

ceunant (Wel.) Evans (1852:191a) gives: "a brook running in a deep channel." Generally speaking, it's a term applied to watercourses running in a deep gorge, hollow or ravine. One can be seen on OSX 186 at SN5441 and the term exists as an element in a number of PN's shown at NGR SH5361, SH9909 and SJ2314. See ceuffos.

chad (E) An agricultural drainage term. Wright (1880:295b) gives: "A small trench for draining land. Midl. C." Baker (1854:1:105) states: "A small narrow trench for draining land. In some places the *first spit* only, whether of turf or soil, is termed the *chad*, in others the last spit. This word is, I believe, peculiar to the Midland district. I am inclined to think it merely signifies a narrow trench, without reference to any specific part, and the succeeding word strengthens the conjecture." The foregoing suggests that chad is found in midland counties only, in particular, Northants.

chahen rit (Co.) Borlase (438c) gives: "khahen-ryd, a torrent." Lhyud's form (165a) is: "kahen-ryd", also 'torrent' and Pryce has: "chahen rit, a land flood, a torrent." Williams (1865:41b) spells it: "cahenryd" and states further that: "this word is only found in the Cornish Vocabulary, where it is written *chahen rit*, torrens." This must refer to the Borlase (421b) incorrect spelling "chabenrit", corrected, in the later Norris Cornish Vocabulary list (337) as: "chahen rit". The foregoing suggests, not so much a permanent watercourse, more likely a term used in times of flood. See keynres.

chanáil (Ir.) See canal.

chanel (ME & OF) See channel.

channel (E & Sc.) From the OF *chanel*, 'canal' and variously defined as 'rivulet, stream, the bed of a river, watercourse in a street, gutter', Palsgrave (1852) gives: "the broke or chenell, *le ruisseau*." The ME form is chanel. In Sc. and OSc., not so much a watercourse, moreover, a street gutter. There are a number of mapped examples, two are shown at NGR TA2621 and TQ9424, and SK1904, shows a flood relief channel. See the MED.

chenell (E) See channel.

cinder drain (E) An agricultural drainage term; an alternative term for a gravel drain, used by Loudon (708:4288).

cladha(i)n (SGael.) Dwelly (202a) defines this and cladhan as: "channel, very shallow stream."

clais (Ir? SGael. & Wel.) In Ir., this term is questionable – furrow, pit or trench being the common definition given in most Irish dictionaries, Dineen (1904:144a) is the exception, coming nearest with: "clais(e), a drain, sewer." For the SGael., Dwelly (203a) gives clais(e) as: "furrow, gutter, ... ditch", however, shown on OSX 404 at NJ2006 and OSX 450 at

ND1936 are two instances of 'clais' as watercourses, so, in Scotland at least, it has obviously become a generic term in a transferred sense. In Wel., more than one dictionary gives a watercourse definition; GPC, Pugh and Spurrell's all list clais as 'rivulet, trench' or the like, but, unlike the Sc., map references are lacking.

claise (Ir? SGael?) See clais.

clash (Mx?) The Mx. form of the Ir. clais. Kelly (45a) defines it: "a furrow, trench, gully." See clais.

clay pipe drain (E) An agricultural drainage term. Loudon (709) gives: "The *earth drain*, called also the *clay-pipe drain*." Johnstone (129) quotes an authority that has used the clay pipe drain: "this clay pipe has conducted a small rill of water a considerable way under ground, for more than twenty years, without any sign of failing." However, it's not one of the common terms to be found in drain terminology. See earth drain.

cleuch (Sc.) See cleugh.

cleugh (E & Sc.) The Sc, form of the E clough. Heslop (162) gives: "a dell, or cleft through which water runs" and Brocket (1:96): "cleugh" and "clough." A cleugh, in Nrth., is shown on OSOL 42 at NY7694 and the Sc. forms, cleuch and cleugh can be seen on OSX 321 at NX9592 and OSX 345 at NT6968 respectively. This term is commonly found as an element in PN's; Catcleugh (NT7403) for one. See PNNorDu (41); EPNE (1:99) and clough.

cleush (NIr.) Patterson (20) gives: "Cleush, a sluice; a water channel or spout."

*clóh (OE) See clough.

clote (ME) A drainage term. See EPNE (1:100) and PNC (315-6) for its use in FN's.

clough (E) From a hypothetical OE **clóh*. The PP (88) gives: "Clowys", water schedynge: Sinoglositorium". Goodall (104) states: "This characteristic word comes from OE *cloh*, a ravine with steep sides, usually forming the bed of a stream or river." One can be seen on OSX 268 at SJ9668. See EPNE (1:99) and cleugh.

clowys (ME) See clough.

cocker (E) Parish (29) gives: "A culvert; a drain under a road or gate." The EDD (1:696) gives: "coker".

cockey? (E) A watercourse or drain? A term apparently peculiar to Norwich; Marshall (1787:377) gives: "The grate over a common sewer, hence, probably, Cockey-lane, in Norwich." No reference here then as a natural watercourse. See cokeyam; the MED under cokei; The EDD (1:686b); Norwich (1892:103a) and, for an extended discussion, PNNf (1:5-7).

coegnant (Wel.) This compounded term is listed in the GPC (531b) defined as: "small stream", in use since 1795.

coileach (SGael.) Dwelly (223b) defines this and its dim. variants coileachain and coileachan as: "rill, rivulet."

coileacha(i)n (SGael.) See coileach.

coker (E) See cocker.

cokeyam (L) The L form of the E cockey. Norwich (1:358, note 16) tells us that: "Obstupavit Cokeyam. Cockey is a local word now used for a gutter. In the 13th C there were several in the city which were sufficiently permanent watercourses to be used as abuttals of lands." The term is mentioned again in (1:226 & 2:16). See cockey.

colimbus (L) WW give three different equivalent OE terms from the various vocabularies: (1:184.12) "wæterscipes"; (191.5) "wæterþeote" and (211.13) "wætergelæt", collectively, these words define colimbus as some sort of 'aqueduct, conduit or water channel'. This is certainly the meaning given in Late Latin (3rd to the 6th C) for the variant colymbus, a term which is also used in ornithology.

colymbus (L) See colimbus.

condict (Sc.) Jamieson (1:482b) gives: " a conduit."

condie (Sc.) Warrack (100a) gives: "a conduit, a drain." See all con- and cun- entries.

conduct (E) Ogilvie (1:546c) gives: "A channel; a conduit." The term is now superseded by conduit.

conductus (L) Martin (220a) gives: "conduit." See conduit.

conduit (E) From the F *conduit*, and, ultimately, the L *conductum*, *conductus*, 'to lead, to bring'; a term which has a great number of variant and corresponding forms throughout the ME period up to ModE. Ogilvie (1:547a) gives: "A pipe, tube, or other channel for the conveyance of water or other fluid". WW (1:564.47) give: *"condyt"* and (1:733.40): *"cundyth undyr the erthe"* from a *Nominale of the 15th Century*. An early pluralized form, *condwys*, appears in the *The Ayenbite of Inwyt*, published in 1340. Carr (97) gives: "cundith" and The SLB (1906:245): "cundytt". Most conduits were used for the supply of water rather than as independent watercourses. There are many conduits shown near Oxenhope on OSOL 21 at SE0234, some of which were used to supply reservoirs. Some other forms which can be found in glossaries are cundy, cundie and cundeth. See conduct; cundard; cundiff; cwndid; the Sc. conduit terms, condie; cuddie; cundit and cundy; the MED and OED for lists of forms; PNBrk (859); PNDb (3:721); PNGl (4:113); PNLei (1:28,108, 210) and PNYW (7:173).

condwys (ME) See conduit.

condyt (ME) See conduit.

coorse (Mx.) Kelly (51a) gives: "a course, ushtey, a watercourse."

corafon(ig) (Wel.) Evans, Pughe & Pryse and the GPC all give this term as 'rivulet, brook' and corafonig, its dim., is given by Evans (1:191a) as: "brooklet."

cornant (Wel.) Evans, Pughe & Pryse and the GPC all list this common element, variously defined, as: 'brook, brooklet, rivulet, streamlet' etc. Lhyud (165a) gives the unusual form: "korrnant" and the term is used at least once as a PN element, in Pantycornant , shown at NGR SS9688.

counter drain (E) An agricultural drainage term. Gwilt (957) gives: "A drain parallel to a canal or embanked water-course, for collecting the soakage water by the side of the canal or embankment to a culvert or arched drain under the canal, by which it is conveyed to a lower level." Ogilvie (1:604c) gives much the same. A counter drain can be seen at NGR TF1719 running alongside the R Glen near Spalding, in Lincs.

counter-trench (E) See trench-royal.

course (E & Sc.) Adopted from the F *cours*, 'a natural or artificial channel for water'. Turner (1880:426) has "cowrses", the ME form. A mapped example can be seen on OSX 173 at TQ3399, and a Sc. one on OSX 321 at NX8384. More often found compounded as water-course, mill-course etc; Stephens (1848:96) gives: "mill-course or rivulet."

cove (E) The meaning of cove in the English Lake district is: "a steep sided valley" or "a recess in the side of a fell", Prevost (55) etc., but OSOL 6 at NY2206 shows Little Narrow Cove, rising east of Scafell Pike, as a watercourse generic.

cover (Co.) There is a cofer fros mentioned in BCS 1056 (S684) – here fros is the generic and cofer (cover?) the name. In St. Agnes CP, near to this stream, is Gover Farm which may be the present form of the OE 'cofer'. Bond gives: "Brook – cover, gover." See go-; gu-; fros and gofer.

covered-drain (E) Rees (33) and Stephens (1848:96) both mention this agricultural drainage term.

covered gutter (E) An agricultural drainage term. Elworthy (1886:162) gives: "A drain made with square sides and flat top and bottom." See culbit.

cowrse (ME) See course.

creek (E & Ir.) This term has its roots in the F *crique* and Dut. *kreke*. The MED list creek under crike but it's not a strong ME contender as a watercourse generic term, moreover, it favours an estuarial one. Ogilvie (1.622a) informs us that this word is used: "rarely in England." Rare that is, as a watercourse, but used it is; quite a number of creeks are shown on OS maps, as inland watercourses, at NGR TF4430, TF5401 and TF6529. On OSX 249 creeks are common west of the River Nene outfall to The Wash. Although creek is mainly used around the coastal areas of Great Britain in a different sense, some parts of Lincs. have obviously adopted the term as a watercourse generic. In Ireland, creek is also used of an inland watercourse – OSIDS 64 at R3755 shows one draining into the R Shannon. In Bel. and Gmy., kreek and Kreck, respectively, are used as a stream term.

creignant (Wel.) See crygnant.

crigyll (Wel.) Pughe & Pryse (1:367a) give: "a ravine, a creek" and the R Afon Crigyll is shown at NGR SH3274.

crike (ME) See creek.

crook (Sc.) Jamieson (1:535b & 542b) gives: "crooks" and "crukis", repectively, "the windings of a river." Cramalt Crook is shown on OSX 336 at NT1825, so here, crook has been appended as a generic term.

cross drain (E) An agricultural drainage term mentioned by Stephens (1848:72). Two are shown on maps, at NGR TF1612 & 1909.

cross-furrow (E) An agricultural drainage term. Ogilvie (1:631a) states: "A furrow or trench cut across other furrows, to intercept the water which runs along them, in order to convey it to the margin of the field."

cross gutter (E) An agricultural drainage term. Elworthy (1886:306), under gutter, mentions this term in the following statement: "You 'ont make thick field dry 'thout some cross gutters."

crukis (Sc.) See crook.

crundel? (E & OE) This term is questionable – KCD (3:xxi) states: "This obscure word seems to denote a sort of watercourse, a meadow through which a steam flows." B-T (Supp., 135) agrees. Morris (1857:39) gives: "a watercourse" and Duignan (1905:48) states: "Crundel is an A-S. word the meaning of which has long been doubtful; it is now settled as 'a ravine, a strip of covert dividing open country, always in a dip, usually with running water,' EDD. The word is found in over sixty charters, on manorial boundaries." Earle (1888:383) (S 564) gives: "mæres crundel" and "mæres crundelle". As mentioned above by Duignan, there are over sixty instances of crundel in charter bounds but, topographicals need checking! In some instances 'watercourse' definitions could be acceptable but not all charter bounds refer to valleys. Although crundel is disputable as a qualifier, evidence suggests that there is good reason to include it here, and, although now fairly antiquitous compared with contemporary PN etymologists, the work of F H Baring (300-3), who discusses the Brks. crundels in depth, is worth perusing, so too is Grundy (47-50) who extends the discussion with regard to Hants. There is a Crundalls Farm, in Worcs., shown at NGR SO7876 and another in Kent, shown at TQ6642. See PNBrk (773) and, on the bounds, (861); B-T (172) for a list of variants and The EDD (1:826a).

crygnant (Wel.) Morgan (91 & 128) offers two place names in the form "crynant." The GPC form is "creignant" and a crygnant can be seen on OSX 215 at SJ0400.

crynafon (Wel.) Obviously a dim. compound – cryn plus afon – therefore, 'small river or stream', Spurrell's (97b) give "brook, stream."

crynant (Wel.) See crygnant.

*crype (OE) See crypel.

crypel (OE?) This is a doubtful contender – B-T (Supp., 135) give: "crypel, a narrow passage, burrow, drain". See EPNE (1:118, under *crype); PND (244); PNSx (1:16) and PNWo (21, under Cripplegate) for its connection with grype, 'furrow, drain'.

cuddie (Sc.) Jamieson (Supp., 89a) gives: "A ditch or cutting to lead the drainage of a district to a river; also, an overflow connection between a canal and a river." See all con- and cun-entries.

culbert (E) See culvert.

culbit (E) Elworthy (1886:174) gives: "culvert called also a *barrel arch*, that is, a circular conduit made of brick-work." See covered gutter.

culvert (E) Brogden and Wright both give: "drain"; Peacock (1889:151) gives: "CULBERT – a culvert; an underground tunnel for conveying water." See cundiff.

cundard (Co.) A cundard is a "conduit" according to Thomas (79).

cundeth (ME) See conduit.

cundie (ME) See conduit.

cundiff (E) Peacock (1889:151) gives: "CUNDIFF, CUNLIFF. A culvert or conduit, an underground tunnel for conveying water", and Brockett (51), gives: Cundy and Cunliff as: "a conduit." WW (1:733.40 and 800.24) give the ME forms: *"cundyth(e)".* See conduit.

cundit(e) (ME & OSc.) The OED1 (2:793c) gives: "cundite" and The DSL (DOST) gives: "cundit" as "a channel for water." See all con- and cun- entries; the MED, under conduit and PNNt (176, 262).

cundith (ME) See conduit.

cundy (E & Sc.) Brockett (51) gives: "Cundy ... a conduit." Heslop (210) renders cundy: "a drain, a sewer, a conduit", and mentions that: "A rummelin' *cundy* is a drain with loose, broken stone laid round to allow of percolation from the surface." For the Sc., Warrack (118a) gives: "a small drain crossing a roadway." See all con- and cun- entries and The EDD (1:839b) for variants and dialectal spread.

cundy-hole (Sc.) Warrack (118a) gives: "a conduit, as one across a road." See all con- and cun- entries.

cundyth(e) (ME) See cundiff.

cundytt (ME) See conduit.

cunette (E?) From the F and listed by Knight (657b) as: "a small ditch in the middle of a dry ditch, to drain the water off the place."

cuniculum (L) WW (1:216.15) give: "greop." See grip.

cunliff (E) See cundiff.

currel (E) A rare term, found only in Suffolk, Wright (1880: 366a) gives: "a rill, or drain, *east*." Rye (54) follows Moor (96) who gives: "a rill or drain", and advances an etymology stating: "Compounded probably of current and rill. *Drindle* is nearly the same – and is also the bed of such a *currl*, or a small furrow." Further references are lacking but it's interesting to find that, in Alb., *curril*, a very similar word, also means 'stream'. A word like curril could have been carried here by some ancient tribe but evidence for this is lacking. Any hypothesis to establish the correct etymology is always worth pursuing but, in the case of *curril*, nothing seems to exist outside of Albanian!

curril (Alb.) See currel.

currl (E) See currel.

cut (E, ME, NIr. & Sc.) Commonly used as a Black Country dialect word for a canal in the Midlands. R. B. Peacock (1869:23a) gives: "a canal or artificial watercourse." Another Peacock, Edward (1889:153) gives: "A drain for draining land, not a sewer; commonly, though not always, one newly made." In addition, Cole's definition (35) is slightly different: "One of the many words for Dyke or Drain, a channel cut for water." A NIr. 'cut' is shown on OSNIDS 20 at J1069 and, for the Sc., Warrack (120a) gives: "an artificial watercourse." One of the longest 'cuts' to be seen on any map runs between NGR SU9178, River Thames confluence and SU8871, near to Ascot Heath racecourse, where it rises. Two cuts are shown on Sc. maps, one on OSX 341 at NS2374 and another on OSX 448 at NC6948, which, more than likely, is the most northerly. See the PNGl (4:117); PNLei (1:225-6); PNWe (2:245); PNYW (7:176) and *cut, for a different sense.

*cut (ME) The common occurrence of Cut(t)mills, Cuttemills and Cutel or Cuttle Mills, together with cut- prefixed watercourses, listed in PN books, suggests that the Wallenburg comments (144-5): "a *"cut"* or artificial channel carrying water to the wheel" is very much the likely derivation, and on the same page, under Allington, he also mentions a lost "Cuddymill" and states that: "The K pl-n provides as a matter of fact the earliest evidence of the word *cut*. It is not found in OE", however, he also mentions that "Its early appearance in Kent, the pl-n material of which is practically free from Scand elements, is a point strongly in favour of native origin." Conway (29) also mentions the Cuddymill and gives: "Cuciddemille" from a 12th C document and AD (3:171) give: "… lands in Westram, called Cuttmyll …." In BCS 922 (S 1577), boundaries of land relating to BCS 921, is a *cutelwlle* (sic) which may be genuine. There is a Cutmill in Dor., shown at NGR ST7716, lying just north of Sturminster Newton. A

Cutmill and a Cutwell Hill, in Gloucs., is listed in PNGl (3.169) and (1.110) respectively. There is a Cuttle Mill in Northants., PNNth (103-4) and a Cutte Mill listed in PNO (125). The PNSr (77) list yet another Cut Mill, which is shown at NGR SU9145 and the PNWa (45) have Cuttle Mill and (205) Cuttmill, which may relate to the entry in the AD (3:274) which states: "a moiety of two watermills in Greneborwe, called 'Cuttole'...." Finally, we have another Cutmill listed in the PNWo (91-2). Cut-, as a prefix in watercourse names, is well evidenced too – PNDb (664) list Cuttle Brook, shown at NGR SK3728; other Cuttle Brooks are given in PNNth (103-4); PNO (7) and PNWa (2). In addition, there is a Cuttail Brook in PNNt (3) and a Cutsyke PN listed in PNYW (2:70-1), which, according to OSM, has a small stream or cut close by. Most of these names parallel the G stream names given by Jellinghaus (142) who gives: "Küttel-Bach" and "Kutelbeke". The Kuttelbach, shown on Google Maps, may refer to the same place. Förstemann, in his *Altdeutsches Namenbuch* (2:1766a) under "Cuttelbeke", gives some variant forms: "Kottelbeke; Kötelbecke and Küötelbieke"; all stream names, so, bearing in mind these very relative corresponding names, the point regarding native origin, mentioned above by Wallenburg, would suggest that we should at least acknowledge the possibility that, our own cut(t)- names, are based on an OE *cut- or *cuttel element that has somehow come down to us from the Gm., therefore, further research will be needed to clarify the situation fully. See cut for a different sense; PNBrk (862); EPNE (1:120-1); PNO (7 & 439); PNYW (7:176) and all the relative PN books mentioned above.

*cutte (ME) See *cut.

*cut(t)el (ME) See *cut.

cutting (E) Warwickslade Cutting is a drainage ditch shown on OSOL 22 at SU2705, a photograph of it can be seen on Geograph. Dictionaries do not define cutting as a watercourse so the mapped and photographic examples are, therefore, the only references.

cweorn-burna (OE) See mill-burn.

cwndid (Wel.) The Wel. form of the E conduit.

cwter (Wel.) The Wel. form of the E gutter.

cwyrn-burne (OE) See mill-burn.

cwysig (Wel.) Pughe & Pryse (1:385b) give: "a small furrow, a drain."

D

dæl (OE) See dale.

dælf (OE) See delf.

dal (Got.) See Balg (65a) and dale.

dale (E) From the OE *dæl* and Got *dal*. The EDD (2:14a) gives: "A river-valley between ranges of hills or moorland", which is much the same as that given by Atkinson (1868:134-5) who mentions dale as a term for: "The distinctive name of valleys which run far up between the high moorlands of Cleveland ... with a small rapid stream or beck". Clarke (304) also mentions dale as a Lincs. drainage term. At least two dales can be seen on the modern map, both as watercourses – one on OSX 279 at SK6696 and another on OSX 281 at TF0790. See EPNE (1:125-6).

dam (E & Sc.) Although dam is not usually defined as a watercourse there are at least three dams to be seen on OS maps, shown at NGR SE5755, SE6032 and TG3615. The DSL (DOST & SND) do not give a watercourse definition for dam – Sc. or otherwise and the nearest Jamieson (1880:2:10b) gets is "Improperly used to denote what is otherwise called a mill-lade, Kinross". Further, dam is listed in Warrack (124a) where here, it *is* defined as "a mill lade". This term differs from the SGael. *dàm* which always includes a grave à.

dàm (SGael.) There are two 'An Dàm' watercourses on the OSX 360 map (2002 edn.). One runs between Loch na Sailm and Loch a' Chreachain at NGR NM8814 and the other runs into Loch Avich at NGR NM9012. Dwelly (309a) gives 'conduit' as one of the definitions for dàm; an, in SGael., is an obsolete word for water, the current meaning is '*the*', therefore, the name An Dàm could be translated 'water conduit' or 'the conduit' – the latter being more likely. A photograph of the Loch Avich, An Dàm, can be seen on Geograph, It is obvious that the evidence here suggests that dàm is a generic term for a watercourse, confirmed as such, by the modern map.

darent (Co.) Bond gives this as one of a number of watercourse terms.

ddwr (Wel.) See dŵr and llif ddwfr.

deak (E) See dike.

dean (E & O & ModSc.) From the OE *denu* – 'valley', a later form of dene and normally defined as a 'wooded vale of a rivulet' but also applied to the watercourse running through it; names denoting the valley are very often transfered to the stream name. The term is well evidenced on maps – shown at NGR NU0338, 0336 and 2124. There is also a cluster of dean term and brook names around Great Gransden in Hunts., shown at NGR TL2756. For the Sc., Jamieson (2:27a) and The DSL (DOST, under dene) give the same definition as the E. A Sc. dean is shown on OSOL 16 at NT8525. See den; EPNE (130) and ERN (119).

deek (E) See dike.

deep (Sc.) Jamieson (2:33b) gives: "The channel or deepest part of a river." In Dut. and Fle., diep is used of a 'stream, waterway or channel'.

defending-trench (E) See trench-royal.

defer (E & OE?) a variant of devr and dever, 'water', suffixed to RN's such as Candover. See dobhar.

deke-holl (E) Rye (58) gives: "Deke-holl, Dike-holl, A hollow or dry ditch. Not necessarily a dry ditch." Wright (385a) gives: "DICK-HOLL."

delf (E & ME) From the OE *dælf*, B-T (194). Cole (37) gives: "DELPH , or DELF, One of the many words for a drain or Dyke, a channel delved or dug to carry off water." Peacock (1899:162) has the same. Bailey has an interesting entry in his 1726 dictionary: "KINGS-delf, great ditch which King *Canutus* digged in Huntingdonshire." Three delphs can be seen centred on OSX272 at TF0471. See the MED under delf.

delph (E) See delf.

den (Sc.) Jamieson (2:27a) gives: "DEAN, DEN, A hollow where the ground slopes on both sides; generally, such as one as has a rivulet running through it." A number of dens are shown on maps – OSX 389 at NO5955; OSX 389 at NO6155 and OSX 426 at NJ8748. See dean, and for confirmation of den as a watercourse, King (2008:104-5).

dene (E & Sc.) See dean.

denu (OE) See dean.

deur-rid (Bret.) See red.

dever (Co.) A variant from the Brit. **dubro*, **dubra*, 'water'. As a simplex, defined as 'water', when suffixed by other elements, 'watercourse'. The terminal of the R Anton, in Hants., is given as -dever and -deuer by Leland (1:269) and Holinshed (1:95) respectively. See defer; devr; dobhar and ERN (15-6).

deverhens (Co.) See deverhent.

deverhent (Co.) Bond gives this as one of a number of watercourse terms. Nance (1990:42a) gives: "deverhens", 'watercourse'. The suffix may be connected with the E term *henting*. Ellis (3:52) gives: "henting or water-thorough", if so, then dever, 'water' and hent, 'furrow'. In some Wel. dictionaries, the compound dyfr-hynt is defined as 'water' and 'course'.

deveron (Co.) From the stem dever- one of a number of watercourse terms given by Bond.

devr (Co.) A variant from the Brit. **dubro*, **dubra*, 'water'. This stem prefixes a number of watercourse terms – i.e. *deverhent, deveron* etc. The R Candover, in Hants., is suffixed defer in BCS charters. See defer; dever; dobhar; and for discussions on this stem and its extentions, CPNE (82); PNW (6) and ERN (68-9) for the RN Candover, in Hants, which was Candevr in *The Book of Fees*, 1208-13 (1:47).

dhree(a)n (E) See drain.

dhutelet (E) Robinson (1876a:53) gives: "an outlet or

watercourse."

dib (E) Possibly a variant of dip or dub, 'a depression, a small hollow, a valley'. A dib is shown on OSX 286 at NGR SD4223 as a watercourse. See PNYW (7:181).

dic (OE) See dike.

dick(e) (E) See dike.

dick-holl (E) See deke-holl.

diep (Dut & Fle.) See deep.

dig (SGael.) Dwelly (334b) gives: "ditch, drain, furrow." Three are shown on OSX 390 at NM7167, others on OSX 359 at NM7410 and OSX 454 at NF7168.

dike (E & Sc.) From the OE *dic*, cognate with OHG, *tîh**; NHG, *Deich* and *Teich*; OFris, *dîk*; ON *dîki* and OSax, **dîk*. The Gm. forms are **dîka-* and **dîkaʒ* and the IG is **dʰeigʷ-*, all from Köbler. B-T (203) give: "*a ditch, the excavation or trench made by throwing out the earth, a channel for water.*" The most common forms used today are dike and dyke, especially on maps – dike on OSOL 26 at SE4978 and dyke on OSX 179 at SO8320. Cole (41) states: "the regular word for a ditch." Parish & Shaw give the forms: deek (41) and dick (43) and Wright (376a) gives: "deak." Blaeu's map of Ely to Downham Market, dated 1645, gives: "dicke." The term is now commonly applied to any watercourse, especially in the north. Although the Sc. form dyke is normally defined as a ditch, an excavation or wall and such like, a watercourse is shown on Orkney on OSX 463 at HY3415. See The EDD (2:71).

dike-gutter (E) Dinsdale (35) gives: "DIKE-GUTTER, A ditch running along the bottom of a hedge." Dickinson (108) gives: "Dyke gutter"

dike-holl (E) See deke-holl.

dike-sheugh? (Sc.) Warrack (133a) gives: "a narrow trench or ditch alongside a 'dike'." See dyke-slouch.

dimble (E) A variant of dumble – Evans (1881:138) gives: "a dingle, dell." and "On the N.W. side of the county, the general pron. is 'dumble' as in Derbyshire." The EDD (2:75b) gives: "A ravine with a watercourse through it." See dumble.

dingle (E) A rare term for a watercourse but one is shown, as such, on OSX 266 at NGR SJ3362.

ditch (E, Sc. & Wel.) From the OE. *díc* B-T (203). WW (1:585.19) give: "*Fossa*, a dyche", a ME form, which suggests 'ditch' rather than 'dike'. The usual definition, but not always, is: 'a trench for draining land'; ditch is related to dike, a weakening of it, in the opinion of some and is commonly found in the eastern counties of Eng. One of the longest ditches to be seen on any map is Cat Ditch, shown on OSX 193 running from TL2337 to TL2934. Sc. ones are shown on OSX 330 at NT0213 and OSX 438 at NH7070 – Wel. on OSX 165 at SS7886 and OSX 240 at SJ2610. See SED (118).

ditch-trow (E) See hedge-trow.

dob (Ir. & SGael.) O'Reilly (193a) gives "a river, a stream". For the SGael., Dwelly (345b) lists dòb as "gutter", and dob and doib as "river, stream." See doub and dub(s).

dobha(i)r (SGael.) The definition of dobhar, given by Macleod & Dewar (244) is: "water". Lhuyd gives the same. Dwelly's form (345) is: "dobhair". For the Ir., Cormac (40) gives: "dobur". Most authorities listing dobhar include dur also. Aberchalder, in Scot., is usually etymologised to aber, 'mouth of' and dobhar 'water'. The def-; dev-; dob-; dou-; dow-; dur-; dwf-; dwr-; dyf- and dyr- stems are all synonymous and have commonly been defined as 'river' or 'water' or suchlike in the majority of cases. The dobur and dúr terms are also listed in the (CL:EOI).

dobhrág (SGael.) The definition of dobhar is 'water', the logical extension for the dim. would be –an, which would give dobh(a)ran – 'streamlet' and the like. Liddall (22), under Douranside, mentions that: "Douran is for Dobharan a dim. of dobhar, water." CM (520) under Aldourie, by the author A. M., states: "I have been much puzzled with this name, until I was told that the burn or "Allt," which enters Loch-Ness at the place, is called in Gaelic the "Dourag," or the little river, from "Dur," in Gaelic, and in Irish, water ; the "ag" expressing the dim.." MacBain (1922:140) gives: "Aldourie, from the

"Dourag" burn, while Dourag itself is from dobhar, water."
On Aldourie, see Ross p 7. The dobhran and dobhrag forms
are dims. formed from the same stem; *ag* is found in Irish as a
dim. termination, Watson (1911:364) mentions: "dobhrag",
and (364) states that: "Parallel to dobhrag is dobhran." There is
an Allt Dobhrain shown at NGR NN2741 and Bundoran, in
County Donegal, Ireland is *Bun Dobhrain*, in Gaelic – both root
back to dobhran. See Index (E) under d(h)obhar, dobhrág and
dobhrán for a huge number of variant forms of this very
widespread element, which can all be found in CPNS.

dobhrán (SGael.) See dobhrág.

dobur (Ir.) See dobha(i)r.

dock (E) The only evidence for this term comes from OSX 226
at TL5377 and OSM, where Braham Dock (drain) is shown as
a watercourse entering the River Great Ouse, below Ely.

dofer (Co.) From the Brit. **dubro*, **dubra*, 'water'. Only Dexter
(1926:37.92) gives: "dofer; douer; dour; dower and dur" as
"water, river" most others give 'water' only. See dobhar and
dur.

doib (SGael.) See dob.

doke (E) This term suggests a connection with the ON *dokk*,
'hollow, valley' etc, Charnock (1880:11) gives: "a stream, the
sike or syke of the northern counties." Kennet (11b) has much
the same and Louly (5) defines it: "a small brook." See PNWe
(2:246).

dokk (ON) See doke.

dol (Co.) Given in Williams (1865:107b) corresponding with the
E dale, 'a valley through which a stream runs', but unlike the E
term, it does not appear to have become a generic in a
transferred sense. See Nance (1978:234b).

dolek (SNn.) Jacobsen (1928:110b) gives: "a small watercourse,
brook." See dullack.

doriov (Gyp.) Gypsy words are difficult to come by; this one is
defined as "river", by Smart & Crofton (75). See doyav.

dotarchlais (Ir.) Only in O'Reilly (198b) as "watercourse".
Lhuyd (354b) gives: "a conduit pipe."

dothar (Ir.) Lhuyd (354b) and O'Brien (185a) list this term as "river", O'Reilly (198b) gives the same and its variant "dothuar". For the SGael., Dwelly (355b) gives: "dothuar, for dobhar obsolete word for river."

dothar-chlais (SGael.) Dwelly (355b) defines this compound as: "conduit, water pipe, channel."

dothuar (Ir. & SGael.) See dothar.

doub (Ir.) Windisch (501b) gives: "doub fluss; dob river, stream." See dob.

double-trench (E) See trench-royal.

douer (Co.) A variant of dour. Dexter (1926:37.92) gives: "water, river" and Norris (354) has L "aquam", 'water', see dofer.

douet (A-N) Moisy (327) gives douet and douit as: "petite cours d'eau, ruisseau." and (333) the variant duit. Eyton (375-80), quoting a L deed c1200 also gives duit proving A-N usage in Shropshire. The E form is dowt, given by Wright (1880:401a) and, in particular, Brogden (57) as: "a dyke, a ditch, a drain." See The AND for other forms.

douit (A-N) See douet.

dour (Co.) From the Brit. *dubro, *dubra, 'water'. Dexter (1926:37.92) and Norris (354) give: "water or river." most authorities agree. There are a number of variant forms, Pryce gives: "thour, a river, a brook." The R Dour, in Kent, Dover and all other Dov- PN's are all rooted to *dubro. See all the do-entries and dur; Pokorny (264); PNRB (341) and, for dour used as a common generic appended to RN's, Weatherhill (1-3).

dourag (SGael.) See dobhrág.

dower (Co.) A variant of dour. Dexter (1926:37.92) gives: "water, river", also the second element of the PN Ruthdower. See dofer.

dowr (Co.) A variant of dour. For a great number of Co. RN's, using dowr as a common generic, see GKN. George (1998:32b & 2009:171a) lists this word too, as "river". It is also used in a number of compounds, see dour; Nance (1978:235a & 1990:42a); the list in Bond and George (1998:33a & 2009:172a).

dowrbons (Co.) Nance (1978:235a) and George (2009:171b)

both define this term as: 'aqueduct'. GKN gives two forms for aqueduct – dowrpons and ponsdowr, however, in the case of dowrbons, the references do not give any instances of a wider generic status for aqueduct, unlike the E and Sc., who do.

dowrhens (Co.) One of two compounds from the stem dowr- given by Bond and confirmed in Nance (1978:235a). George (1998:33a & 2009:172a) has the corresponding "dowrhyns, watercourse". See dever and dever-.

dowrhyns (Co.) See dowrhens.

dowrpons (Co.) See dowrbons.

dowt (E) See douet.

doyav (Gyp.) This and doriov are listed under river by Smart & Crofton (183).

drain (E, NIr., Sc. & Wel.) From the OE *dreahnian*, B-T (210).The most common agricultural drainage term found with many compounds (eighty plus!) and dialectal variants. Ross (1877:52) gives: "dhreen" and "dhreean"; Parish & Shaw (47) "drean"; Long (18) "drine" and Moor (113) "dreen". There are many other variants. At Lothal, in India, a 'drain' was discovered, during excavations of the site, dating back to 2900 BCE and some sites go back even further. See OSX 280 at SK8298 for an E drain; OSNIDS 4 at C6933 and OSNIDS 20 at J1067 for NIr.; OSX 330 at NT0921 for Sc., and OSX 164 at SN4101 for Wel.

drainer (E & Sc.) Oldfield (6) states: "... an old gote and drainer, called Symond's gote ...." For the Sc. Ogilvie (2:96a) states, amongst other things, that a drainer is: "A stream from a lake, morass, &c.; as, the Leven is the drainer of Loch Lomond."

drain-well (E) Knight (741a) states: "A pit sunk through an impervious stratum of earth to reach a pervious stratum and form a means of drainage for surface water ...."

drama (L) See drana.

drana (L) Martin (233a) gives: "drain, a watercourse." Bailey gives: "DRAMA, a drain or water-course."

drang? (Co.) According to Thomas (81) this term signifies: "An open drain or gutter; an open groove or channel." Jago

(1882:155) has much the same: "... trench, gutter, or drain." The term does not appear in later Cornish dictionaries.

draucht (Sc.) The Sc. form of the E draught with very few watercourse references; one such can be seen in The DSL (SND) given in the form: "drawght". See watter drawcht.

draught (E) A watercourse bearing this appellative is shown on OSX 286 at SD4740. On the OS First Series map, Sheet 44 dated 1828 is a place in Glous. called Draught, but there is no stream course shown. See draucht; wydraught and PNGl (1:131).

draw-dyk (Sc.) Only in The DSL (SND) as: "A ditch serving to draw off water."

drawght (Sc.) See draucht.

drawn (E) Dartnell (47) states: "In a water-meadow, the large open main drain which carries the water back to the river, after it has passed through the various carriages and trenches."

dreahnian (OE) See drain.

drean (E) See drain.

dreen (E) See drain.

dribble (E) Only defined in the compound dribble drain and not as a simplex.

dribble drain (E) An agricultural drainage term. Apparently, 'a tributary drain'. Arkell (325) gives this term in his paper, *On the Drainage of Land*, which is also reported in the BCAG (86) stating that: "Stone drains are various; the most common here are wall and dribble, or rubble, the former as main, the latter as tributary." It's a term that starts a sequence of alternative terms by various authorities giving alternative names to a particular drain prefix – dribble or rubble; rubble or wall; wall (or box) leading to 'sough' (Loudon) or triangular; triangular or wedge; wedge (or plug) or brick - the list just goes on and on! There is not always too much to choose between them as they are all agricultural drainage terms built exclusively for the purpose of draining land. See rubble-; stone-; wall- and all the other relative terms mentioned and cross references therein.

drift (Co. & E) For the Co., Jago (1882:156) renders this term:

"A trench cut in the ground resembling a channel dug to convey water to a mill-wheel." In E, apparently, a watercourse associated with mining; one which may not necessarily run on the surface. Knight (749b) defines it as: "a passage in a mine ... or a drain for carrying off the water." However, OSX 332 at NU0603 does show a surface 'drift', north of Rothbury, in Nrth.

drift-hole (E) This is a rare term, perhaps used only in Lincs. Peacock (1889:179) defines it as: "An underground channel for conveying water from one drain to another." See drift.

dright (Sc.) Only in The DSL (SND) as: "A mill-race."

drill (E & Sc.) A 'brooklet, a streamlet of the smallest kind; a mere trickle'. Ogilvie (2.101b) gives: "a small stream; a rill." The Sc. connection comes from DSC (2.469b) defined, amongst other things, as: "a small dribbling brook: caochan, alldan." George Sandys, in his *Poetical Works*, (1:cii) also uses the term: "... from hence in smaller drills her course she keeps...." seconded, in (2.347) with: "Whose drills our plants with moisture feed."

drillock (E) This term comes from Robertson (39) defined as: "A gutter by the roadside."

drindle (E) Rye (64) gives: "A small channel to carry off water; a very neat dim. of drain", and Moor (114) states: "A small slow run of water".

drine (E) See drain.

drock (E) A dialect form of the OE *proc* and *purruc*, surviving in some counties. Dartnell (48) gives "A short drain under a roadway, often made with a hollow tree", and from other quotes gives: "Where meaning a water way, it is usually spoken of as a drockway, "drock" alone being the passage over the ditch ...." Robertson (40) gives "Drock, The same as DRUFF", defined on the same page as: "a covered drain, generally one built of rough masonry." See druff; EPNE (2:213-4 & 217); Forward (59 & 295); PNGl (4:181) and PNW (339-40).

drockway (E) See drock.

droke (E) Possibly a west country form of drock – Jago

(1887:65b) lists droke, as well as trone and vore, under furrow. Thomas (82) gives, "a slight channel". As trone is "a trench or drain" in Dev., Marshall (1873b:75), so too, perhaps, droke. One of the early forms of Drockmill Hill in Sx. (PNSx 446) has droke as a first element. Further evidence is lacking, making the droke term questionable, on the other hand, see trone, and Dartmoor (in bibliography).

drove (E) Ogilvie (2:105a) states that, in agriculture, a drove is: "a narrow channel or drain, much used in the irrigation of land." Smiles (1867:1:67) also mentions the term.

drowning-carriage (E) Dartnell (48) gives: "a large watercourse for drowning a meadow."

druff (E) Robertson (40) gives: "A covered drain, generally one built of rough masonry." The OS First Series 1828 map, Sheet 44 gives: "Througham or Druffham". See drock; PNGl (1:120) for a dialectal PN form, and BCS 180 (S259) which gives: "*Ðruhham*".

drumba (E) See dumble.

drumble (E) See dumble.

druve? (E) Wright (408b); Halliwell (321a); Grose (52) and Lously (8) are the sources for this term given as: "a muddy river", so, without further elucidation, it must be considered doubtful as a qualifier.

dub(s) (E & Sc.) Apparently related to the Got. *diups*, 'deep', Balg (71-2). A term found in the E Lake District and Scot. Brockett (1.147) describes dub as: "a small pool of water; a piece of deep and smooth water in a rapid river. Celt, *dubh* a canal or gutter." Most glossaries and dictionaries do not give dub as a watercourse generic but a number of dubs can be seen on maps, one is shown on OSOL 4 at NY0944 and another at NY1647. The plural dubs is shown on OSOL 6 at NY2103. Jamieson (2.119a) gives, amongst other things: "A gutter." For a Scottish example of dub see OSX 319 at NX4874. Further useful references can be found in Prevost (64a) and, especially, PNCu (1.5) for confirmation of dub as a watercourse. See also dob; Pokorny (267-8) and The DSL (DOST & SND).

*dubra (Brit.) See dofer, dour and dŵr.

*dubro (Brit.) The term is also listed in the (CL:EPC). See dofer, dour and dŵr.

duct (E) The OED1 (3:702b) cites an Act of 1766 and gives: "For making and perfecting any channel, course, main cut, or duct, through any of the grounds."

duit (A-N) See douet.

dullack (Sc.) Possibly from "dolek, a small watercourse; brook", given by Jakobsen (1928:1:110b). See The DSL (SND).

dumble (E) A watercourse common in Notts. Two dumbles are shown on OSX 270, centred on SK6961, and there are two lying SW of Southwell at SK6752 and SK6850. Holland (108) gives the similar words: "DRUMBLE or DRUMBA, a small ravine ... having a little stream or *rundle* at the bottom", to which dumble is related. See dimble; EPNE (1.137) and, for an extended discussion on dumble, PNNt (279-80).

dur (Co.) A variant of dour. Dexter (1926:37.92) gives: "water, river". See dobhar, dofer and dyr.

dur red (Bret.) See red.

dwfr (Wel.) See dŵr.

dŵr (Wel.) From the Brit. **dubro*, **dubra*, 'water'. Pughe & Pryse (1:643a) give this and its variant dwfr as: "a stream"; GPC, "water" only. A dŵr can be seen on OSX 214 at SN97880 and another on OSX 256 at SJ1357. A much rarer form, ddwr, is suffixed to a stream name shown on OSX 188 at SN9048. See dour and all the dyf(f)- and dyfr- stems.

dyche (ME) See ditch.

dyffryd (Wel.) See dyffrydan.

dyffrydan (Wel.) This is given by Evans (2:886a & b) and Pughe & Pryse (1:656b) as: "streamlet." The terminal -an suggests a dim. form as in ffrydan. However, dyffryd, as a generic does not appear to be supported other than its showing as a Shropshire PN element at NGR SJ2919, together with Little Dyffrydd, nearby, at SJ2920.

dyffrynt (Wel.) Morgan (26-7) states: "In the ancient Welsh laws the word *dyffrynt* is used to denote a river." See dyfr-hynt.

dyfnant (Wel.) See dyfrnant.

dyfrbistyll (Wel.) This term is listed in the GPC (1125c) defined as: "water-spout; conduit, water-course." in use since 1567.

dyfrbont (Wel.) This term is listed in the GPC (1125c) defined as: "aqueduct", in use since 1858. See dyfrffordd, pont and traphont.

dyfrffordd (Wel.) This term is listed in the GPC (1126a) defined as: "aqueduct", in use since 1809. See dyfrbont, pont and traphont.

dyfrffos (Wel.) Lhyud (69a) spells it "dyvrffos" – Pughe & Pryse (1:242b, under awell) give: "aqueduct", confirmed by the GPC (1126a) and the compound dyfrffos melin, is given by Evans (2:341b, under mill-leat and millrace).

dyfrffos melin (Wel.) See dyfrffos.

dyfrglawdd (Wel.) Pughe & Pryse (1:655a) give: "a water trench, a dyke."

dyfr-hynt (Wel.) Evans (2:1858:1063b) and Pughe & Pryse (1:644b; 655a) both list this term as: "watercourse." See deverhent.

dyfrl(l)e (Wel.) Evans, Pughe & Pryse, Spurrells and the GPC all give this as: 'watercourse, bed of a river', or the like. The double 'll' form, dyfrlle, is listed in Spurrells (167a) and the GPC (1127a).

dyfrnant (Wel.) Evans (1852:191a) renders this compound: "brook." Morgan (46) gives: "Dyfnant – A compound of dyfn, deep, and nant, a brook" – a PN in Powys, formerly Breconshire. Three watercourses bearing the dyfnant term can be seen on OSX 187 at SN8458-59; however, the dyfrnant form would appear to be unmapped.

dyke (E & Sc.) See dike.

dyke gutter (E) See dike-gutter.

dyke-slouch (Sc.) Warrack (152b) gives: "a ditch or open drain at the bottom of a 'dyke'." See dyke-sheugh.

dyling (E) See dything.

dyr (OGaul.) Chalmers states that: "dyr", in ancient Gaul., means "a water, a river", being a variation of dur." See dobhar.

dything (E) Only in Brogden (60) given as: "A small drain cut for drainage purposes." Thompson (704) gives: "Dyling – A small excavation for drainage", which may be a provincial variant of dything.

dyvrffos (Wel.) See dyfrffos.

E

e (ME; OE & Sum.) This word-letter, a variant of ea, is found in the ASC (Thorpe, 1:170a), 'river'. There is a R 'E' shown at NGR NH5414, which can claim to be the shortest RN on record in Scot., ('O' Brook on Dartmoor is the shortest in Eng.). The Sum. is given by Waddell (68a) and Prince (92) as a "variant of *a*, water", confirmed by Halloran (3). See a; aa; ae; ea; ee; the MED and The OED1, where it is given as a cognate of ae.

ea (E & OE) From the OE *éa*, B-T (223) give: "*Running water, a stream, river, water.*" BCS (258) gives: "*Meonea*" from a charter dated c790; it also features in the ASC for the year 893. WW (1:177.36) give: "*fluuius*, singalflowende ea" and it's listed in Skinner as: "fluviolus". Peacock (1869:27b) gives: "A river … Kent, as it runs down the Lancaster Sands, Morecambe Bay, is called "The Ea". Ekwall, in his PNLa (190) gives: "Ay" as a variant of the R Eea, which: "points to ON *á*, "river" as the source rather than OE *ea*." There is so much that can be written about this term that it's difficult to know when to stop. For its size, this two-lettered word has probably produced the greatest number of forms, of any watercourse term, ranging from the ones found in the earliest OE charters, through the L and ME periods, before finally arriving in the ModE period with its many dialectal variants. No fewer than fifty different forms can be gleaned from the vast number of documents, maps, charters, word books and the like, presented to us over many years, some of which are just simply amazing. We have a-; e-; h-; i-; o-; re- and y- stems, all of which have their roots in this simple two-lettered word. In addition, the Gm. languages, which have many variants too, are all cognate with the E forms. Ea, just happens to be the name of the Mesopotamian God of Water – so, is it a coincidence, or a fact, that the ea form was carried much the same as the Sumerian *'a'* and *'e'* was carried, through the Asian and Middle Eastern countries, to the shores of the Indo-Germanic ones? There is also a R Ea in the Basque country and an Owenea in County Donegal, Ireland! There does not appear to be an example of ea on the modern map; the nearest is eea, shown on OSOL 6 at SD2164. However, old maps such as Blaeu's map of Downham Market to Kings Lynn 1645, *does* give the ea form. See EDD (2:223b); EPNE (1:42-3) for lists of associated PN's and the numerous listings in the EPNS volumes; especially PNSa (1:333) for Yeaton, who's DB name was Aitone, the ai prefix being a variant; ERN (138) and Embleton (1-4), who has a lengthy discussion on 'a', 'ea' and 'eau', worth perusing.

eaa (E) A variant of ea. Ellwood (19) gives: "Channel of a stream … The Leven and Crake are thus at times called *ea* or stream." WW (1:546.17) equate *eaa* with amnis. See a; aa; ae and ea.

ea-coorse (E) Robinson (1876a:58b) gives: "Ea-coorse, or Eau-course, the water-channel."

eæ (OE) A variant of ea. Wells uses this form often in *The History of the Drainage of the Great Level of the Fens* (5, 6, 7, 16 etc.) and Earle (1865:94) gives: "eæ" from the ASC. BM (2:18b) mentions: "Elme Eae" and a Little Eae is also shown on old maps. See a; aa; ae and ea.

eágor-streám (OE) B-T (244) give: "*A water-stream, water*" and égor-streám in the same sense.

eah (OE) BCS 348 gives: "*grafon eah*". Middendorf (45) gives this form too. See ea.

eáh-streám (OE) B–T (226) give: "a water-stream."

eá-lád (OE) B–T (227) give: "A water-way."

earace (OE) Earle (488a) gives: "*earace*, watercourse." The term is also given in KCD 1064 (S352), in a charter dated 881AD. See racu.

eá-riþ (OE) B–T (233) give: "A water-stream."

earth-drain (E) An agricultural drainage term. Ogilvie (2:96a) mentions this term as does Loudon (709) who adds: "earth drain called also clay pipe drain". Rees (34) gives: "hollow earth-drain".

eas (SGael.) Dwelly (385b) gives eas, easa and easan all as: "stream with high precipitous banks otherwise waterfall, cataract." All the terms are supported with map references – eas at NGR NN3501 and NN3504, easa at NG7829 and the dim. easan at NR7980 and NR8565.

easa(n) (SGael.) See eas.

eá-streám (OE) B-T (236) give: "*A water-stream, a river.*"

eau (E) Adopted from the F for 'water'; a variant of ea. Thompson (705) gives: "a drain." Peacock (1889:188) states that: "EAU" is "a river which falls into the Trent" and Cole (42) gives: "A watercourse." The term has been influenced by the F eau, 'water', and, as a generic, can be seen on OSX 272 at TF0875. The variant, howe, is shown on OSX 295 at TA1054. It is also common on OSX 283 and there is a R Eau shown at NGR SK9098. See a; aa; ae; ea; howe and ERN (140).

eau-course (E) See ea-coorse.

eay (E) A variant of ea. Sternburg (34) gives: "A pond or pool; also a drain or artificial water-course." See a; aa; ae and ea.

édre (OE) See ǽd(d)re.

ee (OE) A variant of ea. Willis (viii) gives: "ee", "ree" and "rhee", and, (349) "Le Ee." The AD (6:455) mentions: "a river called Pever Ee …." Norwich (1:321) gives: "the fishery called Trous Ee", alternatively given in note 2, as: "Trous water." There is a R 'Ee' near Dockum, in Holland and, in G, Ee is still used of a channel. See a; aa; ae; ea and the MED under e.

eea (E) A variant of ea, shown on OSOL 6 at SD2164 and also at SD3778.

égor-streám (OE) See eágor-streám.

ég-streám (OE) B-T (244) give: "*A water-stream, a river*" and éh-streám in the same sense.

éh-streám (OE) See ég-streám.

ei (OE) Earle (1865:76) gives this rarity: "Tinan Þære ei"; it's a variant of ea.

eia (OE) Lhuyd (290a) gives: "eia" as a variant of ea.

eie (ME) See ey(e); ea etc.

eileach na mulne (SGael.) Dwelly (390a) gives: "the channel bringing water to the mill." The term can be seen on OSX 402 at NN6496 and Index (E) gives a number of variant forms.

ek (Co.) See ick.

*er- (IG). See rithe.

*ere (IG) See runnel.

esc (Brit. & OGael.) See ascaig.

escaig (SGael.) See ascaig.

esk (Brit. & OGael.) See ascaig.

esklets (E) Robinson (1876a:60-1) gives: "The inland feeders of the R Esk."

étang (F) See stank.

everlasting-trench (E) See trench-royal.

ew (OCelt?) Kennedy (181) gives: "Ea, Ey, Ew and Aw as Old Celtic." Chalmers (44), probably following Kennedy, the same. See ea.

ey(e) (E & ME) A variant of ea – Holinshed (1:171) gives: "where a riueret called little Eie meeteth Withall …", and, on the same page, "another arme called Sheepes eie …." Norwich (1:46-7) refers to: "the outer part of the stream called Trowys Eye" and Holland (116) states: "At Chester we find the "Roodee" and the "Earl's Eye." We have a brook called the "Peover Eye" which seems to suggest that eye is a synonym of a brook." Oldfield (63) mentions: "from the west end of the Eye …" and John Speed's maps have "Eie Flud" in Leics. and "Little Ey" in Rut. Blaeu's map of Downham Market to Kings Lynn, 1645, gives: "The Maids Eye." The Ches. Eye is shown at NGR SJ7872 and the Eye Brook and R Eye in Leics. and Rut., are shown at NGR SK7703 and SK8118 respectively. In addition, there is an Old Eye shown on OSX 290 at SE5226; a R Eye which runs through Upper Slaughter in Glous. and an Eye Water in Scot., shown at NGR NT9362. See ea; ERN

(157); PNLei (2:188, under Eye Kettleby) and (4:74); PNRu (1) for a good list of old forms and Ross (83, under Eyemouth).

eylebourn (E) See bourn and nailbourn.

F

fadhail (SGael.) An estuarial term, Dwelly gives: (1904:401b) "hollow in the sand, formed by and retaining water, after the egress of the tide." McAlpine (1866:124b) has the variant: "faodhail ... a river through a strand" and OSX 455 at NG0697 shows one.

fairguni (Got.) See Balg (85a) and firgen-streám.

falloch (SGael?) The River Falloch, which flows into Loch Lomond, is far removed from Muckle and Little Falloch, in Angus (formerly Forfarshire), shown on OSX 389 at NO4074 and 4174 respectively. The fact that a river and two other streams carry the falloch name suggests that there may be good reason to consider falloch as a transferred stream term, but, there is very little evidence to support this view which must be left open to question.

faodhail (SGael.) See fadhail.

fead (Ir.) Joyce (1869:3:110 & 229) gives two instances of this term used in PN's. See also (1869:1:458-9, under feadan).

feadain (SGael.) See feadan.

feadan (Ir., Mx. & SGael.) The dim. of fead – O'Reilly (638a) gives "a brook, runnel, streamlet". Joyce explains it in (1869:1:458-9) giving; "Fiddaunnageeroge" in Mayo (shown on OSIDS 23 at F9811) as a PN which uses the fiddaun form and the same map shows at least 45 fiddaun prefixed stream names! The Mx. form, given by Kelly (80a) is feddan, defined as: "a pipe, a whistle" and not as a generic term. For the SGael., Dwelly (420a) also gives the variant feadain and defines both as: "canal, water-pipe" only. However, at least three instances of feadan are shown on Scottish maps as watercourses at NGR NB3335, NG0587 and NN1795. The IOMS shows the PN Ballaneddin at SC3593 (for which Kneen states: "Feddan, a whistle, pipe, etc., in place-names means a narrow stream running in a deep channel like a pipe.") and Boalniddan at SC2068. Both places are discussed by Broderick (44a & 71b). See fead.

feddan (Mx.) See feadan.

feeder (E, Sc. & Wel.) A term used for watercourses which supply a constant flow of water to replenish that lost in canals through lockage, commonly found on maps such as the Stratford-upon-Avon canal feeder, which runs from Earlswood Lakes at NGR SP1174. Phillips (1803:317) refers to the River Witham streams as feeders: "No water to be taken from the river Witham feeders." There are four feeders shown on OSX 206, a Scottish one on OSX 342 at NS6672 and a Wel. one on OSX 240 at SJ2420. See feeder-stream and PNGl (3.96) for another type of feeder.

feeder-stream (Sc.) The common term for watercourses, which supply canals, is feeder; feeder-stream is uncommon. Wilson (1855:1:48) mentions the term but it's not clear from the reading if it refers to a canal feeder or a feeder supplying a pond or lake. Nevertheless, it still adds one more to the ever growing list of -stream compounds.

feith (Ir. & SGael.) O'Reilly (640b) gives: "a boggy stream ...." Joyce (1869:2:397) agrees and for the SGael., Dwelly (428a) gives: "bog channel." Hogan (498b) gives: feth and feith and OSX 443 has six examples of fèith with others showing at NGR NH9532 and NJ0035. Further references are given in CPNS (439 & 454).

fèithean (SGael.) This term can be seen on OSX 457 at NB2725, It is the dim. of feith – Dwelly (428a) gives: "bog channel."

fence ditch (E) An agricultural drainage term given by Boswell (38) who states: "a fence ditch, being the bounds of the meadow on that side, and used also as a drain to convey the water into the tail drain." Loudon (726:4414) gives: "This is generally found in one of the fence ditches; for which reason a fence ditch is mostly used, at once fencing the meadow and draining it" and (729:4438) gives: "... by means of these fence ditches the water is discharged into the river."

fence-drain (E) An agricultural drainage term which can be found in Marshall (1787:170) "... the rivulet and the fence-drains ...." and (280) "... opens the fence-drains ...."

fendeek(e) (E) See fen-dyke.

fender (E) Harrison (1898:85-6) states: "This is the old name of two streams in Wirral, upon the northernmore of which the Ordnance Survey, for some reason, has bestowed the new appellation of Birket. The meaning of 'Fender' here is rather obscure ... Seeing, however, that the Wirral farmers seem to call any kind of large ditch or drain a fender, it is not improbable that the term is of comparatively modern origin, and was primarily used for the reason that the streams were fenders, i.e. protectors, against inundations, which were formerly much more prevalent in Wirral than at the present day." The confluence of the current Rivers Birket and Fender are shown at NGR SJ2791, however, over time these names have interchanged – the R Birket on the OS 1840, Sheet 79 NE – Denbigh map, is marked three times as The Main Fender, which can be traced to the eastern side of West Kirby, and the current R Fender has replaced the Ford Brook, which rose near Barnston. See PNCh (1:15 & 23) for a commentary on both names.

fendike (E) See fen-dyke.

fen-dyke (E) Rye (57) states that: "Deeke is very often used for the ditch and bank together, but a fen deeke has in general no bank". Further (72), he gives "Fendeek. A dyke or drain". Thompson (9) mentions that: "Salmon, in *The New Survey of England*" states that: "Cardyke signifies no more than *fen-dyke*." See the BL for: "A Chart of the Fens between Lynn Regis, and Denver Sluice and Wisbich", showing a "Fendike."

feor (Ir.) O'Reilly (238b) defines this and its dim. feoran as: "a rivulet, stream, sewer." There is a prefixed feor- RN shown on OSNIDS 26 at G8316.

feoran (Ir.) See feor.

feth (Ir.) See feith.

ffloode (ME) See flood.

ffordd ddŵr (Wel.) This compound is listed in the GPC (1305b) defined as: "water-channel", dating from 1867.

ffos (Wel.) There is no real difference between the E and Wel. definition of (f)fos(s) – 'a passage for water, canal, channel, ditch or trench'. On OSX 200 there are three watercourses centered on SN8567 and PN's are shown at NGR SN6867 and SS8791. The ffos is also found prefixed and suffixed in a number of watercourse related compounds, see ceuffos; dyfrffos; fos; ffos melin; ffos wast and gwyffos.

ffos melin (Wel.) Evans (2:341b) gives "mill leat."

ffos wast (Wel.) The GPC (1307a) gives: "channel which carries away the water turning a mill wheel." The second element no doubt indicating 'waste' water that has been used.

ffrut (Wel.) See ffrwd.

ffrwd (Wel.) All Wel. dictionaries list this term variously defined as: 'rill, rivulet, stream and watercourse' – Lhuyd (2c) gives the rare: "frûd". Morgan (62, under Aberhonddu) states: "Many Welsh streams and lakes received their names from the peculiar hue of their respective waters, such as Gwenffrwd, white stream", see below. There are a number of PN's, with the ffrwd element as a simplex, shown at NGR SH4556, SN4512, SN5350 and SN6127, also, we have the tautological Ffrwd Brook shown at SO3410. The old form of ffrwd is frut and ffrut, the former has many entries in the LL: "Guenfrut" (222) later Gwenffrwd is now Whitebrook, a stream and village in Monmouthshire and the latter is given in Skene (126). In addition, the camfrut given in the LL charter bounds (140 line 28), which joins the Afon Llwchwr (River Loughor) at SN5906, is the same stream now named as Camffrwd, which is shown on OSX 165 at SN6105. The ffrwd term, and its variant ffryd, can be prefixed or suffixed in many ways; some of which form watercourse compounds such as ffrwd dwfr, 'watercourse'; ffrwd fechan, 'streamlet'; ffrwd lif, 'stream' and ffrwd melin, 'millstream'. Further, it is interesting to note here that the E form of the Afon Llwchwr, *Lough-* is the same form as found in the Irish names of lakes – Lough Allen, -Erne and -Neagh. Skeat (1888:342a) actually gives lough as the Irish spelling of lake, which also enters into the lists of O'Reilly and Dineen, which makes the E form appear much closer to Ir. than it does to Wel. On the other hand, most authorities state that it is nothing more than a loan word from the Wel. See frut and ffryd and on the lough family of forms PNWe (1:209); PNSa (1:184-5) and compare Bret. *louch*; Cor. *luh*; Mx. *logh*; Sc. *loch* and the forms listed in the MED.

ffrwd dwfr (Wel.) See ffrwd.

ffrwd fechan (Wel.) See ffrwd.

ffrwd lif (Wel.) See ffrwd.

ffrwd melin (Wel.) See ffrwd.

ffryd (Wel?) This stem is a variant of ffrwd and always seems to be given compounded, as a dim. term and not generally listed as a simplex; Pryse, in particular, questions its singularity (1866:447b, under cyffryd). There are at least four compounds – ffrydan, is defined by Evans (2:711a, 886a & b) as: "rivulet", "stream" and "a small stream", respectively, to which the definitions given by Pughe & Pryse correspond; ffryden, is defined by Pughe & Pryse (1:50b) as: "A rivulet, a rill, a brook"; ffrydle is defined by Evans, (1:398b) as: "watercourse" and ffrydlif is defined by Pughe & Pryse (2:60a) as: "a stream, a torrent." The GPC gives much the same. See ffrwd,

ffrydan (Wel.) See ffryd.

ffryden (Wel.) See ffryd

ffrydle (Wel.) See ffryd.

ffrydlif (Wel.) See ffryd.

fiddaun (Ir.) See feadan.

field drain (E) Loudon (747a) States: "... field drains are four feet wide at the top, one foot at the bottom, and four feet and a half deep," George Orwell mentions the term field-drain, in chapter 5 of his book, *Animal Farm*.

field-gutter (E) An agricultural drainage term. Lawson (1884b:17) defines grip as: "a field-gutter."

fild-burne (OE) B-T (Supp., 217) give: "*A stream in a plain (?)*."

filum aque (L) See fullum aquæ.

firgend-streám (OE) See firgen-streám.

firgen-streám (OE) From the Got., Balg (85a) gives: "fairguni, mountain" and "firgen-stream, mountain-stream". B-T (288) give: "*a mountain stream*"; and the variants: (288) "firgend-streám"; (289); "firigend-streám" and (353); "fyrgen-streám."

firigend-stream (OE) See firgen-streám.

fl (E) See flood.

flash (E) A watercourse – shown on OSX286 at SD3836.

flat stone drain (E) Stephens (1848:125) mentions this agricultural drainage term and gives a sketch of one.

fleam (E) This term is a later variant of the ME flum. Heslop (809) gives: "FLEAM, a watercourse." Evans (1881:150) "Fleam, a 'mill-tail'; the stream that flows from a watermill after having turned the wheel" and Wright (459b) gives: "FLEME. (A.-N.) A river. A large trench cut for draining. West." Jackson (153) gives: "FLEM, a mill-stream, ie. the channel of water from the main stream to the mill, below which the streams reunite" and further states that "Flem is a corrupt form of flum, an old word found in the early writers." Confirmation of this comes from *The Ormulum* (line 191) which gives: "flumm". The Wright and Jackson forms given above are ME forms listed in the MED. A Mill Fleam can be seen on OSL 128 at SK2328. See flum.

fleet (E & G) From the OE *fleót*, variously, fliét and fleóte, in B-T (292) corresponding with OFris. *flet*; Dut. *vliet*; G *Flett*; OIc. and ON *fljot* and OSax. *fliot*. The Gm. forms are **fleuta-*; **fleutam*, and the IG is **pleud-*, (Köbler). Originally defined as: 'an estuary or arm of the sea', a sense given in WW (1:4.18): "*Aestuaria*" and (229.31): "*Estuaria*" and only later applied to running water, streams and suchlike. Wright (459b) gives: "any stream" or "water"; Peacock (1889:209): "a drain". Battle (45 & 56) uses the form flet, appended to names in various documents, as does BM (1:722a) and the AD (3:127) give: "flete". In the Knytlinga Saga, the Humber is referred to as "*fliot*". EPNE (1:176-7) and ERN (158-9) comprehensively cover this term in great detail and mapped examples are shown at NGR TG4312 and NZ4924. The G *Fleet*, *Fleth*, *Flett*, *Fleuth* and Fle. *vliet*, are still used, as generic terms for "stream, ditch" – all listed in the respective USBG volumes: (GFR:iva) and (Bel.:ivb). See The EDD (398a); PNDb (3:727); PNGl (4:126); PNYW (7:187-8) and PNWe (2:251).

flem (E) See fleam.

fleme (ME) See fleam and flum.

fleót(e) (OE) See fleet.

flet (OFris.) See fleet.

flete (ME) See fleet.

Fleth (G) See fleet.

Flett (G) See fleet.

*fleuta(m) (Gm.) See fleet.

Fleuth (G) See fleet.

fleuve (F) See flum(m).

fley (A-N) Kelham (106b) lists this word defined as: "a river."

Fliess (G) See flos(s).

fliét (OE) See fleet.

fliot (OSax.) See fleet.

fliuch (SGael.) This word is all to do with water and wetness. The fact that a stream course exists on the modern map, shown on OSX 420 at NJ3030, suggests that fliuch has become a watercourse generic in a transferred sense.

fljot (ON) See fleet.

float (E) An agricultural irrigation term used for watering grassland given by Marshall (1796a:86): "One small floodgate turns the stream and three or four sluices, or outlets from the brook, feed the floats". The OED1 (4:334a) gives: "backwater or float." from a 1629 quotation.

floating trench (E) An agricultural irrigation term used for watering grassland given by Marshall (1796a:56). See float.

flod (ME, OE, OFris. and OSax.) See flood.

floð (ON) See flood.

*floda(m) (Gm.) See flood.

*flodiz (Gm.) See flood.

*flodu- (Gm.) See flood.

flodus (Got.) See flood.

*floduz (Gm.) See flood.

flom(e) (ME) See flum(m).

flo(o)d (E, ME & OSc.) From the OE *flod(e)*, 'a stream or large river' which glosses "flumen" in The Vespasian Psalter, c825, Grimm (69). PP (165) gives: "ffloode: flumen". Cognates are OLFrk., *fluod*; OFris. and OSax., *flod*; OHG, *fluot*; NHG, *Flut* and ON, *flóð*. The Gm. forms are **floda-*; **flodam*; *flodi*; **flodiz*; *flodu-* and **floduz* and the IG is **pluti-*, all from (Köbler). Bosworth (312a & b) gives: "*flodus*", the Got. form and the OE "*flod*", from Luke (6:49), and, in the Wycliffe and Tyndale versions (313a & b), he gives: "*flood*" and "*fludde*" respectively. The ME form is flod – the Ormulum has "flod" (line 10612) and the compound "waterrflod" (line 17533). John Speed uses flud and the contracted forms, fl and flu on his 17th C maps, and flu is common on old Irish maps too. The OSc. form is flude, defined by The DSL (DOST) as: "A river". A flood relief channel is shown at NGR at SK1904. See water-flod; the PNBrk (868); PNGl (4:126); PNLei (1:148) and the MED under flod.

flood relief channel (E) Self explanatory. See flood.

flos(s) (A-N, E & G) Kelham (106b) gives the A-N forms: "*flos*" and "*flot*, a flood, a river." Carlyle (7.72) states: "there is one dirty stream or floss (*Hünerfliess*, Hen-Floss) which wanders dismally through those recesses," and Ogilvie (2.198) mentions that floss is: "akin to the G flus, floss, a stream," used locally as "a small stream of water." The 'floss', in the title of the novel, 'The Mill on the Floss', is sometimes taken as a generic instead of the proper name. Fliess, Floss and Flotte are still used as a stream generic in Gmy. The River Floss, which is a fictional name, has been equated with the River Trent near Gainsborough in Lincs. – on this see Biggins, The Mill on the Floss map; flouss and USBG (GFR:iva).

flosh (E) There is a flosh in Cumb., shown on OSX 315 NY5068. No dictionary lists the term as a watercourse.

flot (A-N) See flos(s).

flote (ME) See flowt.

flot-gutter (E) Chope (45) states that: "Flot (vlot) is water or liquid manure for irrigation purposes", and, further that: "the gutters or channels for directing the vlot over a field are called vlot–gutters." In Dev. f and v interchange as in foss and voss.

Flotte (G) See flos(s).

flouing (ME) See flowingus.

floum (ME) See flum.

flouss (Sc.) Jamieson (2:259a) gives: "a flood, or stream", allied to the G flus. See floss.

flout (ME) See flowt.

flow-dyke (Sc.) Jamieson (2:260a) gives: "... apparently a small drain for carrying off water, Banffs." Warrack (183b) defines it differently: "Flow-dyke, n. a drain along the banks of a river; a wall or bank to prevent a river from overflowing."

flowingus (ME) Wycliffe (3:303, from Isaiah 44:3) gives this form which is listed in the MED under flouing as a watercourse term.

flowse (E) Dartnell (58) states: "Occasionally also applied to the narrow walled channel between the hatch gate and the pool below."

flowt (E) Halliwell (365a) gives: "Temse flowt" from a Ms. The OED1 (4:350b) lists flowt under flout, as a watercourse and the MED list theirs under flote.

flu(d) (E) See flo(o)d.

fludde (ME) See flo(o)d.

flude (OSc.,) See flo(o)d.

fluens (L) WW (1:240.46) give: "stream".

flum(m) (A-N, E, ME & Sc.) This comes from the L *flumen*, 'river', which became fleam in ModE. For the A-N, Moisy (471) gives: "*flum*" and equates it with the F "*fleuve*", 'river'. Ogilvie (2:300c) gives the E form: "flume" and, for the Sc., Jamieson (2:261a) gives: "flum" relating it to the OF *flum*, 'water, river'; other Sc. references can be found in The DSL. There are many ME forms too, Wycliffe (4:88b) gives: "flom". Ritson (2:10) has: "flome" and Skeat (1886 lines 2595) lists it too. Langtoft (102) gives: "floum" as does Skeat (1886, line 2898) and Wey (7) gives: "flvn". The Ormulum has: "flumm" on lines 10336 and 10342 and Layamon (line 1299) gives: "flum" and "flom" from various manuscripts. Finally, Morris (1874:2:472, line 8186) gives: "flun", from the famous *Cursor Mundi*, a ME poem of nearly 30,000 lines. See fleam, and for a full list of variants, the MED.

flume (E) See fleam and flum(m).

flumen (L) The earliest OE charter, which mentions flumen, is KCD 11, dated 675 (S 1245). WW (1:177.35) give: "*Flumen*, flod, *uel* yrnende ea", 'river', from a *Supplement to Alfric's Vocabulary*, and KCD 715 (3:347), (S 914) gives: "... flumen Tamense...", 'the R Thames'. Incidentally, as a matter of interest, the earliest mention of the R Thames is given by Caesar (100-44 BCE) in his Gallic War's (*De Bello Gallico*) (5:11) as: "Tamesis." For flod see flood.

flun (ME) See flum(m).

fluod (OLFrk.) See flood.

fluot (OHG) See flood.

Flus (G) See flos(s).

flusch (OSc.) See flush.

flush (E & Sc.) Baker (1854:1:248) gives: "The stream from a mill-wheel." For the Sc., Warrack (184b) gives: "a stream; a run of water." The OSc. form is flusch given in The DSL (DOST).

Flut (NHG) See flood.

fluuio (L) See fluvio.

fluuium (L) See fluvium.

fluuius (L) See fluvius.

fluvie (A-N) Moisy (471) gives this form, other references can be seen in The AND.

fluvio (L) BCS 426 gives: "fluvio" and the variant "fluuio", 'river'. Bede (1:89) gives: "Humbrae fluuio" – the 'R Humber'; "fluuio" is also recorded in KCD 241 (2:2), (S 287) and Skinner, under ea, gives: "fluviolus".

fluviolus (L) See fluvio.

fluvium (L) BCS 860 gives: "fluvium", 'river' and KCD 1160 (5:314), (S 537), gives: "fluuium". There are other entries for these forms in both works.

fluvius (L) WW (1:736.20 & 21) give: "*Hic fluvius, Hoc flumen*, a flod", 'a river', from a *Nominale of the15th Century* and (325.33) "Flumen, uel fluuius, flod", from an *Anglo-Saxon Vocabulary of the 11th Century*. In addition, (1:177.36), they give: "*Fluuius*, singalflowende ea"; the OE translation, from the *Supplement to Alfric's Vocabulary*. BCS (317) gives: "fluvius" and in KCD 148 (1:179) (S 57), "fluuius Carent", the 'R Carent' is given.

flvn (ME) See flum(m).

foghlais (SGael.) Watson (1908:149) explains this term together with the "old spelling Foglais" as: "substream" giving Fowlis as a PN example, which is also discussed in CPNS (458).

foglais (SGael.) See foghlais.

folosg (Ir.) O'Reilly (252b) gives: " a moor or mountain brook."

fons (L) WW list this word as "well" but also "burne" in (73.5).

foot-drain (E) An agricultural drainage term. Vancouver (1808:285-6) states: "the next step was to cut foot-drains, or drains one foot wide and one spit deep ...."

foot-trench (E) An agricultural drainage term. Marshall (1796:381) gives: "FOOT Trenches; superficial drains, about a foot wide."

for-bai (ME) see forbay.

forbay (E) Apparently 'a channel between the millrace and the millwheel' Addy (79) gives: "in a deed dated 1630, relating to property at Dore, mention is made of the weirs, forbayes, etc., belonging to a corn mill." See the MED under for-bai and the Durham Roll (3:622 & 916).

ford (E) From the OE *ford*, and found in OHG as *furt*; Frank*furt*, 'the ford of the Franks'. Ogilvie (2:313c) gives: "a stream; a current." The term was first recorded in a stream sense in 1563 – see the OED1.

fors(s) (Sc.) Commonly defined by the majority as 'a waterfall', Jamieson's main definition (2:286a) gives: "a stream, a current."

fos(s) (Co? & E) A Latin survival from fossa, 'ditch, dike, trench'. The Cornish *fos* does not appear to be used in the same sense as the Wel. *ffos*, 'gutter, drain, water channel' or the E. term. Williams (1865:152b) defines it as: "a ditch, a moat, a trench," but with no reference to running water, therefore, a doubtful qualifier in Cornish. In E, the term is not disputed – Bridlington (20) mentions: "... a fosse for bringing water from Ruddestain to Castelburun." A foss is shown on OSX 290 at SE5154, which falls into the R Ouse above York; just about as near as we can get to a L survival on the modern map. "The Fosse (a slow streame yet able to beare a good vessell)", mentioned by Holinshed (1:159) is the R Foss which falls into the R Ouse in the centre of York. In fact there are at least four Foss Rn's in Yorks., all discussed by Ekwall ERN (162-3) and further in PNYE (3-4); PNYN (4) and PNYW (128). Middendorff (53) gives charter evidence for foss in OE and, further (118), equates the Dutch slochter with L fossa. See ffos.

foss(e) (E) See fos.

fossa (L) The root of the Co. fos, E foss and Wel. ffos. WW (1:585.19 & 585.19) give: "dic and dyche", from a *Latin and English Vocabulary of the 15th Century*. See the terms mentioned.

fox (E) Peacock (1889:219) defines this term as: "To carry one drain under another by means of a tunnel of wood or masonry." See wolf.

frafal (ME) The MED give this term as a questionable "tail race (of a mill), the drop below the mill wheel."

frechewater (ME) See freshwater.

french drain (E) A drain filled with stone or gravel; two examples can be seen on Geograph at NGR SP3159 and TQ2072.

fresh (E) Wright (482a) gives: "A little stream or river nigh the sea." See the MED.

freshet (E) A stream that feeds a mill. See The EDD(495b) and Rotzol (112).

freshwater? (ME) WW (1:799.18) give: "*Hec ammis*, An frechewater."

fros (Co.) The earliest reference to this term can be found in an

Anglo-Saxon charter. BCS 1056 (S684) gives: "cofer fros", here, fros is the generic and cofer the name. The cofer may have developed, at some time later, into gover, as in Gover Farm, which is situated near to this boundary brook. The lists of Bond, George and Nance all list this term as "stream". See cover; frot, and, on the early appearance of fros, CPNE (100-1).

frot (Bret. & Co.) The Bret. form, given by Henry (126), is *froud,* which corresponds with the Co., frot – Wel., ffrwd, 'torrent'. Loth (63) gives: "*froud,* courant, ruisseau" and (204) "*frut, frot, frout* – ruisseau, courant". For the Co., Norris, *Cornish Vocabulary* (364) translates frot: "alveus, a channel or strait." Zeuss (2:1119) gives the same. Williams (1865:326a) lists frot under stret. BCS 544 gives a charter instance of frot: "mercfrot", which is rendered "mercfliot" in KCD (1063). According to the VCH H&IOW (434 note 4) "mercfrot" is thought to refer to the Blackwater Brook; in B-T (673b) merc equates with mearc – a word all to do with marks and boundaries, so we should consider, cautiously, that here, frot may mean 'boundary brook', although, as a generic in Co., the definitions of Norris and Zeuss do not really confirm it as such whereas the later work of Nance (1990:59a) does. See ffrwd, fros and CPNE (100-1).

froud (Bret.) See frot.

frout (Bret.) See frot.

frûd (Wel.) See ffrwd.

frut (Bret. & Wel.) For the Bret., see frot; for the Wel., ffrwd.

fry (E) Wright (1880:486b) states: "A drain. Wilts" and Dartnell, in his Wilts. word book (61), gives: "a brushwood drain."

fuaran (SGael.) Dwelly (460a) gives: "well, spring, fountain"; fuara, a contracted form of fuaran, can be seen on OSX 438 at NH6678.

fuaranan (SGael.) The dim. of fuaran, which can be seen on OSX 431 at NH3336.

fullum aquæ (L) Martin (247a) gives: "stream of water", confirmed by Black (529a) who states: "a stream, or stream of water."

fur(r) (Sc.) An agricultural drainage term from the OE *furh*, 'furrow'? If so, then this term is a contraction. The DSL (SND) give: "A deep furrow or rut cut by the plough to act as a drain for surface water", also the variant furr. See furrow.

fur-drain (Sc.) An agricultural drainage term given by Warrack (196b) who states: "a small trench ploughed periodically in the land for drainage." See The DSL (SND) under furr.

furrow (E) From the OE *furh*. An agricultural drainage term given by Dickson (1765:144) who states: "a passage into the furrows or drains."

furrow-channel (E) An agricultural drainage term given by Loudon (1243) in his definition of gutter: "a furrow-channel or drain." It would appear to be the only reference.

furrow-drain (E) An agricultural drainage term given by Ogilvie (2:348) who states: "to drain, as land, by making a drain at each furrow, or between every two ridges."

fyrgen-streám (OE) See firgen-streám.

G

ga (Sc.) An agricultural drainage term given by Warrack (198a) who states: "Ga, a furrow, a drain; a hollow with water springing in it." See ga(a)-fur.

ga(a)-fur (Sc.) An agricultural drainage term given by Warrack (200a) who states: "Ga-fur, a furrow in a field for letting water run off" and Jamieson (2:333b) who gives the variant: "gaa-fur". See ga.

gain (E) This extremely rare term comes from Lawson (1884b:16) who defines it: "A shallow water-course." Further references are lacking.

gais (Ir. & SGael.) O'Reilly (268b & 269a) gives: "gais, stream" and "gaise, small brook" respectively. For the SGael., Dwelly (472b) gives much the same for both forms. The term is also listed in the (CL:EOI).

gaise (Ir. & SGael.) See gais.

gaisidh (Ir.) No doubt a dim. of gais, O'Reilly (269a) gives: "a stream, current."

gall (OSc.) An agricultural drainage term, the earlier form of gaw. See the DSL (DOST).

gang (E, ME, OE, OSc. & Sc.) From the OE *gang*. Ogilvie (2:361c) gives: "the channel of a stream". B-T (226) give the OE compound: "eá-gang, *A water-course*." Wallenburg (507) lists the ME forms: "Ouergange" and "Overgonge". For the Sc., Jamieson (2:347b) gives "The channel of a stream, or course in which it is wont to run; a term still used by old people ..." and Warrack (203a) gives the same. See the DSL (DOST & SND).

gang water (Sc.) Reid (134) gives: "... gang water to a little mill ...."

gaut (E) Atkinson (213) gives: "A narrow opening, whether in a row of houses, or in the soil, sufficing to afford a passage, for men, &c., in the one case, for water in the other. Spelt also Gawt, Gote" and Halliwell (395a) gives: "GAWT. The channel through which water runs from a water-wheel. *Lanc.*"

gaw (E & Sc.) An agricultural drainage term listed by Stephens (1889:1.121) who gives: "Gaws or water-runs should never be neglected to be cut after lea ploughing." The term would appear to be uncommon in England but more widespread in Scotland, both in O and ModSc. Warrack (206a) gives: "Gaw, a channel or furrow for drawing off water; a hollow with water-springs in it." Jamieson (361b) gives much the same. See gaw-cut; -fur; -furrow; water-run and The DSL (SND).

gaw cut (E & Sc.) An agricultural drainage term. Addy (86) gives: "a small, narrow 'grip' to drain hollow places in a corn field into the trenches", and Stephens (5:127) states: "gaw-cuts must be made with the spade in every hollow on the surface and across head-ridges, even on thoroughly drained land to quickly carry off large falls of rain."

gaw-fur (Sc.) An agricultural drainage term. Jamieson (362a) gives: "A furrow for draining off water." See gaw.

gaw-furrow (Sc.) The same as gaw-fur, see The DSL (SND) under gaw.

gawt (E) See gaut.

ge-lád (OE) B-T (Supp., 351) give: "*a lode, water-course.*" See EPNE (2:8-9) and PNBrk (3:754, 886).

geleed (Fle.) See lead.

gel-fur (E) Prevost (79) gives: "A water furrow; a deep furrow made either longitudinally or across the ploughing to carry off excess of water." See gaw; ga-fur and gaw-fur.

ge-ríþe (OE) B-T (Supp., 397) give: "*A small stream, rivulet.*" See ride; rithe and riþig.

ghyll (E) See gill.

gifta (L) CHR (30) gives: "a stream."

gifta aquæ (L) Black (540b) gives: "The stream of water to a mill."

gil (Ice., Mx., ON & SGael.) See gill.

gill (E, Sc. & SGael.) A great number of English word books all have entries for gill – Parish (1888:63) in Kent; Heslop (324) in Northumberland; Ellwood (1895:26) in Lakeland and Robinson (1855:71) in Whitby, just to name a few. Manx, Scottish and Scottish Gaelic dictionaries all have entries too. Although commonly found in the northern counties of England, gill is also found in Kent, Sr. and Sx., as well as the IOM and Scot. (Lowlands to Shetland) and the Gaelic regions therein. The parent word is the ON gil 'a ravine with a stream', which can certainly be applied in most cases with the exception of the southern counties. The Sr. and Sx. word is problematical because of the likelihood of ON 'gil' surviving in these parts despite the fact that gill, as a stream course, is well represented on the modern map. Three gills are shown in Sx. centred on NGR TQ2331, and Sr. gills can be seen at TQ1541 and 2241. In Eng., the two most common forms, gill and ghyll, can be seen on the same map within less than 1 km. of each other just below Langdale Pikes in the Lake District at NY2806 and as many as thirty 'gills' can be found on one of the Yorkshire Dales maps. Broderick (2006:112a) gives a Manx PN example of gil, and, for a Scottish Lowland gill, see OSX 336 at NT0326. The most northerly gills (Shetland) can be seen on OSX 452 at

ND3467, and the SGael. gil and gill, can be found close to each other on the same map at NB5454. The gil form here is identical to the Ice. gil, still in use today as the second element in a vast number of stream names. See IED (199b) and PNYW (193) for a massive list of PN's. The problematic Sr. and Sx. gills are discussed in PNSr 85, 368 and PNSx 203-4 has an extended discussion on the subject.

gill-runnel (E) Robinson (71) gives: "a rivulet or thread of water coursing along a deep dell."

gill-stream (E) Given as "gill-stremes" in the OED1 (4:163a) from *The Wars of Alexander*, by Skeat (1886:line 3231).

gipsey (E) Ross (1877:68) gives: "gipsey a spring of water, issuing from the earth with great force" and Grose (76): "Gypsies, springs that break forth sometimes on the Woulds of Yorkshire". OSX295 at TA0765 shows a gypsey and OSX 301 at TA0272 a gypsey race. See lavant and the EPNE (1:202).

girt (E) The Somerset variant of gurt.

glais(e) (Ir., SGael. & Wel.) This term is found in P- & Q- Celtic, in many different forms. O'Reilly (281b) uses the form "glaise." A glais can be seen on OSIDS 45 at M1829 and a glaise on 64 at R0045. Joyce (1922:102) gives: "glaise, glais and glas, a streamlet." supplemented by the usual anglicized forms – "*glasha, glash, glas and glush*" and PN examples (1869:2:272): "Finglas", (2:67): "Glashagloragh", (2:475): "Glashgarriff" and (3:590): "Tullyglush." Hogan (437b-439a) has four columns devoted to glais. Two glash streams are shown on OSIDS 64 at R2046 & 2049. For the Mx. the IOMS shows a number of instances of glass that have survived as PN's – Ballaglass SC4084, Ballaglass SC4690 and there is a R Glass at SC3584. The most notable though, must be Douglas at SC3775, 'dark stream', See Broderick (28b & 29a). In SGael., the term is well represented – CPNS (456-8) is a must, and gives many, many forms of this widespread element, all listed in Index (E). One of the few mapped instances of 'glas' can be seen at NGR NF7227. The Wel., form glais is listed in Spurrell's (1934:195b) and the GPC – supported by PN evidence in Morgan (134) and DPNW (xlv). The PN can be seen at NGR SN7000. In addition, the RN's Dalch and Dawlish in Dev. belong here too – see PND (4-5) and EPNE (1:203b) for other D- related PN's.

glaisin (Ir. & SGael.) Dim. of glais, 'streamlet'; Joyce (1869:2:71) gives "glaisin" and (1869:3:364) mentions that: "Glasheens in Mayo; [uses the] Engl. plural instead of Irish Glaisini, little rills", (shown on OSIDS 23 at G0513) and further, in (1:456) states: "The dim. Glasheen is also in frequent use as a territorial designation." OSIDS 64 at R1635 shows Glasheennabaultina River, in Limerick, which Joyce (1869:1:202) defines as: "the glasheen or streamlet of the May-day games." The double 's' form, *glaissin*, appears in Earls (64 Line 17). IOMS shows Glashen Farm at SC2870, and, Ballaglashan at SC2573 is Ballaglashen in Broderick (208) who covers the Mx. element and other PN's. CPNE 320 mentions the SGael. form and Ross (2001:127), the associated PN's.

glaissin (Ir.) See glaisin.

glas (AIr. & SGael.) See glaise.

glash(a) (AIr.) See glaise.

glasheen (AIr.) A PN form, see glaisin.

glashen (Mx.) A PN form, see glaisin.

glass (Mx.) See glaise.

glide (E) This poetical rarity comes from Green's *Never to Late* (272), under Capricornus: "*When he that in Eurotas' silver glide Doth bain his tress, beholdeth Capricorn*". Also given and confirmed by the OED1 (4:217a). There is another reference to glide in Bowlker (40) who gives: "the chief haunts of the smaller Greyling are in glides; but the large ones generally resort to deeper water ..." which would seem to infer that the glide is a shallow part of the stream and not necessarily the stream itself.

gloder (SNn.) Jacobsen (1928:235a) gives: "a steep cleft through which a brook runs ... a brook running between steep banks."

glover (SNn.) See gløver.

gløver (SNn.) Jacobsen (1928:243a) gives: "a cleft or deep gully of a stream." There is also a Burn of Glover marked on maps.

glush (AIr.) See glaise.

goah (Bret.) See goeth.

goal (E) See gool(e).

goat (E & Sc.) A variant of gote; Callis (395) gives: "goats". For the Sc., Warrack (219b) gives: "Goat, a drain, ditch, gutter ...". An E goat is shown on OSX 340 at NU1238 and a Sc. one on OSX 420 at NJ3628. See gote and gout.

goaz (Bret.) See goeth.

gobhar (SGael.) According to Dwelly (511b) a "sort of branching river", a term which can be seen on OSX 446 at NC3754.

goeh (Bret.) See goeth.

goer (Bret.) See gouer.

goeren (Bret.) See goveren.

goeth (Co.) George (1998:47a) defines this term as: "stream, watercourse, conduit ....", repeated, much the same, in his latest dictionary (2009:238a). Williams (1865:181a) uses the form goth, defining it as "vein", also stating that: "goth is another form of gwyth". Bond gives: "stream" for goth, as does Nance (1978:248b). The goth form is also cog. with the Bret., *gwaz*, defined by Henry (148) and Bret. L (130) as: "ruisseau." In addition, Bret. L gives the compound *gwaz-dour*, "ruisseau" and Bret V (90a) lists the forms *goah* and *goeh* as: "ruisseau, rivière". In addition, it is interesting to note that the first book to be published in Breton, in 1499, *Le Catholicon* (Catholicon), a dictionary of Bret., F and L words (108) gives the form: *goaz*, "ruisseau, riuulus", F and L respectively. Loth (206) gives three forms: "*goeth, goez, goaz* ". Overall, goeth, goth, gwyth and the Bret. forms goaz, goez and gwaz are all cognate. See gwyth, CPNE (111) and red.

goez (Bret.) See goeth.

gofer (Wel.) Evans (1852:191a) gives: "brook" and (711a) "rivulet." Pughe & Pryse (2:50b) give: "rivulet" also. The term exists as an element in PN's; one is shown at NGR SH8472 and another at SH9777. See gover and guuer.

goferen (Wel.) The dim. of gofer. Evans (191a) gives: "a small brook".

goffrwd (Wel.) Evans (2:886b) gives: "streamlet" the same as Pughe & Pryse (2:94a).

goil (E) See goyle.

goit (E) A variant of gote. Easther (56) uses this form as the pronunciation of the word gote and, further, states that: "This word is always sounded and spelt *goit*, but if properly *gote*, it would still be *goit* in the dialect." See gote.

goitstead (E.) According to Harland (14b) a goitstead is: "an old watercourse." No other authority appears to list the term except for The EDD which is referenced to the above. See goit.

gole (E & OF) Possibly a variant of gool. Parish (49) gives: "A wooden drain pipe. In the north of England the word is used of a small stream." See gool(e); goyle; EPNE (206) under goule for associated PN's and, especially, PNC (327).

golet(te) (ME) See gullet.

golt (Sc.) Jamieson (2:127a) gives: "A drain, ditch."

gonant (Wel.) This compound is listed in the GPC (1459a) defined as: "small stream", quite a late word – in use from the 18th C onwards.

gool(e) (E) From the OF, *gole*, *goule*. OED1 (4:297c and 298a) give the variants: "goole, goule and goal." Not often defined as a watercourse in regional glossaries, Grose, Kennett and Ray all give: "ditch". See goyle; The EDD(2:683b); EPNE (206, under goule) for associated PN's; PNLa. (53, under Gooden) for a detailed discussion and PNYW (2:16).

gooteris (ME) See gutter.

gord (E) Possibly adapted from the F *gourde*. Kennett (14a) states: "Gord of water, is by Gouldman explained to be a narrow stream of water. Gore." Wright (1880:523a) also gives: "A narrow stream of water". The OED1 (4:315b) gives the form: "gourd".

gore (E) See gord.

gort (A-N) Kelham (116a) and Moisy (513) both have references for this term variously defined as 'stream, pool, current' and the like. See gaut.

gota (L) Martin (252a) gives: "gut, a drain." See gote.

*gota (OE) See gote.

gote (E, ME, OSc. & Sc.) From a postulated OE **gota.* Carr (1:193) gives: "A channel of water from a mill-dam." As stated under goit, and repeated here, Easther (56) mentions: "This word is always sounded and spelt *goit*, but if properly *gote*, it would still be *goit* in the dialect." For the ME, the PP (196b) gives: "Gote, or water schedellys". The OED1 (4:312b & c) suggest connection with Dut. gote which is also listed by Kiliani (157a) who gives: "gote, canalis, tubus." Sc. gotes are shown on OSX 341 at NS3058 and OSX 348 at NS5883. See gaut; goit; got; gout; gowt; The DSL (DOST & SND) for Sc. forms and definitions; The PN's of C; Cu; Db; Nt; We; YW, who all have entries for *gota or gote and the MED for other variants.

goth (Co.) See goeth.

gothen (Co.) See gwython.

got(t) (A-N & Sc.) Kelham (116a) gives: "a sluice, drain, or ditch." For the Sc. Jamieson (2:247a) gives: "got" and "gote" as does Warrack (222a) with the addition of gott (222b) all in the same sense. The DSL has many forms and the word is a variant of the E and Sc. gote; goit and goat(e).

gotyr (ME) See gutter.

gouer (Bret. & Co.) For the Bret., Henry (138) defines it: "ruisseau", 'brook, stream', giving also the MBret. form, *gouher.* Bret. L lists two forms of this term – *goer* (117) and *kouer* (182) and for the Co., Borlase (432b) gives: "a brook ... or bog", confirming Lhuyd's entry (204a) as Armoric. See gover.

goueren (Bret.) See goveren.

gouher (MBret.) See gouer.

goule (E & OF) See gool(e).

goulet (OF) See gullet(t).

gourd (E) See gord.

goush (E) Wright (524b) and Halliwell (412a) both give: "A stream."

gout (E) Possibly a variation of gote; adopted from the F *égout*, 'sewer'. Roberson (60) gives: "A covered drain or culvert."

Sternburg (42) has: "gout" and "goat", defined as: "A ditch or drain." and Lawson (1884b:16) states: "GOUT, A watercourse bridged to make a roadway."

gouttière (F) See gutter.

gover (Bret. & Co.) Cog. with the Bret. and Co. *gouer*, Wel. *gofer*; one of the most common g- forms for a 'rivulet or brook', listed in all Cornish dictionaries from Lhuyd's list (1707) to George (2009). The same form is found in the Brittany dialect of Vannes – Bret. V (97a) defines it: "ruisseau, cours d'eau." See some of the other go- entries, some of which have cog. Bret. forms and CPNE (122).

goveren (Co.) Lhuyd (141b) gives this as Armorican. The nearest Bret. L has is *goeren* (117) and *goueren* (119), both translate 'ruisseau', so far, all Breton forms, but Nance (1990:69a) promotes it as a Cornish term for a brooklet, confirming its validity as a qualifier.

goverik (Co.) This and its variant goveryk are dims. of gover – 'brooklet or streamlet', The headword is in George (2009:252a) and the variant in Nance (1978:249a).

goveryk (Co.) See goverik.

gowt (E) Skinner gives: "Gowts." Cole (57): "GOWT, or GOTE, a drain, or channel for water", and Salisbury (15): "gowt". A gowt is shown at NGR TF2224 as the name of a drain. See gote and gout.

goyal (E) See goyle.

goyle (E) This is the Dev. and Som. dialect form of gole or gool. Elworthy (1886:297) states: "A ravine; a deep, sunken, water-worn gully, usually with a running stream down it." Chope (97) gives only: "a ravine, a gully." Pulman (1871:100) has the unusual "gwyle (goil). A gully, a ravine" and Blackmore, (1:48), in his novel *Lorna Doone*, gives: "We were come to a long deep 'goyal', as they call it on Exmoor." A goyle can be seen on OSX 116 at SY2596 and PND (599) discusses the term.

goyt (E) Grose (73) gives: "Goyt, the stream of a water-mill." See goit; gote and on the RN Goyt, PNDb (8) and PNCh (1:27-8).

gozell (E) See guzzle.

gracht (MDut.) See graft.

graff (E) Peacock (1889:244) gives: "GRAFT, GRAFF. A drain, one newly cut". See graft.

graft (E, MDut. & ME) A term adopted from MDut., Brogden (86) gives: "A drain mostly dividing parishes." There is a graft running through the middle of Fishtoft in Lincs., see OSX 261 at TF3541. Kiliani (158b) gives the MDut. forms: "*graft, graue,* and *gracht*". This term may be associated with the Dan. *groft,* 'stream' given in USBG (Den.:va). See EPNE (1:208); PNGl (4: 131) and PNL (4:91).

grain (E & Sc.) From the ON *grein.* For the E, Heslop (339) gives: "a branch of a river", and Dickinson (20a) lists: "Beck grains, where a beck divides into two streams." Three grains can be seen on OSOL 10 centred on SE075675. For the Sc. Warrack (225a & b) gives: "grain, a branch of a river", and: "grane, the fork of a river", respectively. Near to the source of the Water of Allachy, are the East and West Grains which can be seen at NGR NO4887 and four grains are shown on OSX 321 at NX9998. See allach; grainings; EPNE (1:208); the MED, under grein; PNDb (3:682); PNWe (2:256) and PNYW (7:196).

grainings (E) The plural of grain – 'river branches', which can be seen on OSOL 10 at SE0268.

grand carrier (E) An agricultural drainage term, an alternative to a main, given by Loudon (728:4433) "... mains or grand carriers ...." See carrier.

grane (Sc.) See grain.

graue (MDut.) See graft.

gravel drain (E) An agricultural drainage term. One of many - drain terms given by Loudon (708:4288). See cinder drain.

greep (ME & Sc.) Warrack (227b) gives: "a small trench for draining a field", See grep(e).

grep(e) (ME & OE) From the OE *grepe.* Clarke Hall (139a) gives: "land-drain, ditch, furrow". The ME form in the AD is greep, (6:206): "le Greep de la Dene", referred to in PNWo (390). See greep; grip and EPNE (1:209).

griff (E) From the ON *gryfja*, 'pit, hollow'. See grift and OSOL 27 at 8791 for a mapped example.

grift (E) From the ON *gryfja*, 'pit, hollow'. As graff and graft exist side by side; both as drains, so too do griff and grift. Peacock (1889:247) defines it as: "A channel shaped out by water for itself; a runnel." Clarke (313) also mentions grift as one of the terms for a drain. Old maps show Boy Grift near Alford, and Wold Grift near Mablethorpe, two watercourses in Lincs. These names now appear on modern OS maps with drain appended, at NGR TF5078 and TF4882; a good example of how an older name can become modified, sometimes to the point where the original term is lost. Here, it could easily be assumed that 'Boy Grift' is the name, and 'drain' the generic, especially as drain is a very common term, but, in this instance, we have old map evidence to thank for preserving the correct term.

grindle (E) Rye (91) gives: "A small and narrow drain for water. But Drindle is a better word."

grindlet (E) Kennett (14a) gives: "A grip, or ditch; in Southern parts, a grindlet or grippe ...." Grose (74) and Ogilvie (2:430b) both give: "a small ditch or drain".

grip (E & Sc.) From the OE *grípe*, B-T (487) give: "*A ditch, drain*". PP (201) gives: Grype, or a gripyll, qwher water rennyth away in a lond ...." The term may be connected, according to The OED1 (4:428c), with OE *grep(e)*, given by Clarke Hall (139a). Peacock (1889:248) renders it: "A small temporary surface drain" and Atkinson (233): "A trench or furrow hollowed along the surface; a channel or small ditch." The glossaries of Gloucs., Kent., Sx., Whitby and Wilts., all list this very widespread term and Jackson's *Shropshire Word-Book*, (186) cites the ME form: "GRYPPE, or a gryppel, where watur rennythe a-way in a londe, or watur forowe", which comes from the PP not the CA as stated. The OED1 (4:428a) gives the same PP form. For the Sc., Warrack (228a) gives: "a furrow or drain for draining a field." Eng. grips do not appear to be marked on maps which leaves the Sc. ones, shown on OSX 336 at

NT0027; OSX 464 at HY5639 and grip and grips on OSX 464 at HY3949 and 4048, as the only representatives. See grep(e); greep; EPNE (1:211) under grype for FN's and PN's; The EDD (2:730-1) for the dialect regions and variant forms; The OED1 (4:428a); The DSL (SND), under gruip, for more Sc. forms and the MED.

grípe (OE) See grip.

grippel (ME). See gripple.

gripple (E) PP (201) gives this, under grype, as: "gripyll", a ME form. Rye (91) gives: "A small drain, stream, or beck … Surely a diminution of 'grip'." Stating further (93, under grup) that, in determining the hierarchy of the term: "A trench, not amounting in breadth to a ditch. If narrower still, it is a Grip; if extremely narrow, a Gripple." See the MED under grippel.

gripyll (ME). See gripple.

gro (SGael.) This comes from the ON *gróf* 'a stream', found in Yorkshire, England (YE 198). Mackenzie (371) states: "Gro, a stream, is quite common in Lewis, and seems to come from grof, which in the first instance means a pit, ravine, but which has been retained in Lewis in the sense of a stream. We have no less than 63 'gros' in the names of streams in Lewis." Watson (1904:267-8) gives: "Grof, a pit terminally gro, a very common stream ending; probably originally applied to streams which cut their way through peat". However, a good number of 'gro' streams in Lewis, all seem to have qualifying generic terms of their own, such as allt, the gro being compounded with some other element to form a proper name – there are a few exceptions though – such as Isogro, shown at NGR NB4552, and Starragro at NB3848.

grof (ON) See grough.

grøft (Da.) See grough.

groop (E) From the OE *gróp*, '*a ditch, drain*' B-T (488), cognate with OFris., *grope*, as well as other Germanic language forms. Rye (93) gives: "Grup, Groop. A trench, not amounting in breadth to a ditch." Brockett (1:200) lists: "grip, or groop … Also a small ditch, or open drain in a field" and Kennett (14b) has: "Grupe, a ditch, North." See EPNE (1:210); The EDD(2:737b); B-T (488, under grop); all the Köbler dictionaries and the MED under groupe. For PN evidence see PND (1:230); PNCh (2:140) and PNWe (2:257).

groove (E) Not normally defined as a watercourse but is marked as such on OSOL 21 SD9234.

grough (E) Possibly from the ON *grof* or Da. *grøft*. Addy (312), under grift states: "On the Derbyshire moors are little narrow valleys, worn by the water which flows down them, called grufts. The word means a trench, as in MARGERY GREAT GROUGH, p. 144, great there being probably M.E. greot, greet, gret (our grit), sand, grit, or the Old Norse *grjot*, rocks, stones." Two groughs can be seen on maps – one on OSOL 1 at SE0801 and another at NGR SK1498. See PNWe (2:257) and PNYW (7:197).

groupe (ME) See groop.

grove (E) Kennet (14b) gives: "a gripe, grip, or ditch, Lincolnshire" and Brogden (88): "A ditch, dyke, or watercourse." Wright too (1880:535b), also defines this word as: "a ditch or drain".

gruft (E) See grough.

gruip? (OrkNn.) Marwick (82b) gives: "a groove channel cut in the earth." related to gruito perhaps? See grip.

gruito (OrkNn.) Marwick (82b) gives: "a rut through which water can drain away." See grip and gruip.

grup (E) See groop.

grupe (E) See groop.

grut (E) Peacock (1889:249) gives: "A rut, a grip, or small surface-drain."

gryfja (ON) See griff and grift.

gryp(p)e (ME) See grip.

gryppel (ME) See grip and gripple.

gudjil (E) See guzzle.

gue (Sc.) Only used compounded with mill-. See mill-gue.

guerthour (Co.) Pryce gives: "a channel of water." Norris (483) states: "The term guer-thour, 'water-course', seems to be synonymous with ryn ...." and Williams (1865:196b) spells it "gwerdhour", pointing out that the word is a compound of gover and dour. The foregoing, whatever its origins, confirms this term as a watercourse generic.

guher (Co.) Borlase (434a) would seem to be the originator of this gover variant defined as: "river", which is confirmed by Jago (1887:133a).

guitear? (SGael.) Armstrong (756a) gives: "Gutter, Guitear, clais uisge", as clais is defined, amongst other things, as 'trench, furrow, gutter or ditch' and uisge as 'water', we have here, possibly, 'a gutter or ditch of water', but, not necessarily running water, so, that's just about as near as we can get to a generic term for guitear, which leaves it questionable as a qualifier.

gulf (E) Not normally defined as a watercourse but two are marked as estuarial courses on OSX 315 at NY3163.

gull (E) Possibly "a variant of gool" states OED1 (4:503b). Pegge (29) gives: "a deep gutter where water runs" – Rye (93) adds: "a brook thickly overgrown with underwood or brushwood" and Sternberg (44) mentions the term as: "a drain or small stream." A gull is shown on OSX 212 at TM2655.

gulla (L) Martin (254a) gives: "gully, a watercourse." It also appears in the *Coucher book of Selby* (10:xv). The word is the root of the E gullet and OF goulet. See gull.

gullet-hole (E) Dartnell (71) defines this term as: "a large drain-hole through a hedge-bank to carry off water." See gullet.

gullet(t) (E & Sc.) From the L *gul(l)a*; OF *goulet*, 'gully, channel'. The -et(t) suffix suggests a dim. of gull. Wright (1880:539a) gives: "a small stream" as does Peacock (1869:38). Turner (1880:13) gives the ME forms: "golet" and "golette" recorded in 1515. Bailey adds: "also a little stream or accidental course of water." For the Sc., Warrack (233b) gives: "Gullet, Gullot, a water-channel." A gullet is shown on OSX 201 at SO3467. Weld Gullett in Essex appears in a 1659 lease to John Searle, see Battle (216) and PNEss (86). Other gullets are mentioned in PNNth (46 & 165) and PNBrk (872). See gulla.

gulley (E) See gully.

gullot (Sc.) See gullet.

gully (E) From the F *goulet*, generally defined as 'a small stream or brook'. Wright (1880:539b) and Kirby (65) use this form; Baker (1854:1:297) has: "gulley". A gully can be seen on OSX 142 at ST6638.

gulph (E) Alternative name for a girt given in The OED1 (10:40, under yeo): "Girts' or 'gulphs' are names given by the moormen [of Dartmoor] to the long, and sometimes deep, excavations seaming the hill-sides, down which the miners led their stream, generally known as the 'yeo'."

guner (Co.) Another gover variant. The entries in Borlase, Pryce and Jago all come from Lhuyd's entry (141b). Norris (470-3) has some *very* interesting observations on the works of Lhuyd and Pryce, including the guner variant.

gurgis (L) WW (1:736.26) give the ME equivalent: "strem", from a *Nominale of the 15th Century*, and (799.44) the ME: "gotyr" from a *Pictorial Vocabulary of the 15th Century*.

gurt (Co. & E) In Co. gurt is defined by Thomas (90) as: "A shallow ditch or drain." Jago (1882:183) states: "A gutter." The Dev. antiquary George Pulman, in his *Book of the Axe* (20) gives: "... the gurts, as the dykes in the Marsh are called", and has another reference to gurt in his *Rustic Sketches* (100) stating that: "The dykes or drains in Colyford Marsh are called "gurts" – synonymous with "rhines" in the Somersetshire Levels." This defines gurt as a 'drainage channel'. Today, all the watercourses in Colyford and Seaton marshes are marked 'drain'; there is a 'girt' shown on OSOL 9 at SS9143, but not in the usual blue watercourse defining colour, that aside, as a generic, it is not in doubt, and, further, it is not unusual to find shared terms in neighbouring counties as this instance shows. The Somerset variant form is girt, a number of which can be seen on old Somerset maps of Exmoor, such as Red Girts and Smokeham Girts (Ordnance Survey, 1:10560 County Series, 2nd edition (c.1900), Sheet 47, Subsheet 01 & 06, respectively). See gord; gort; gurt gut and gurt-wawder.

gurt gut (E) This term is listed by Parish (52) under his entry for gutterdick, "a small drain." – "Taint no use at all for you to make that 'ere gutterdick, what you wants is a gurt gut." This implies that a gurt gut is somewhat larger than just 'a small drain'. See gurt.

gurt-wawder (E) This extraordinary dialectal compound is explained by Pulman (1871:100) as: "(great-water). A river proper, as distinguished from its tributaries, which are all Little-wawders." This is a very rare term, probably the only instance of it to be found in print. Further, he does not independently list wawder as a dialect form of water, neither does the EDD, although it is known to exist. See gurt.

gushill (E) See guzzle.

gustrill (E) See guzzle.

gut (E) Parish (51) gives: "An underground drain for water" – Wright (1880:54a): "a watercourse which empties itself into the sea" and Heslop (349): "a narrow stream." Leland (1:3 & 37) also mentions gut and the term is common on the Pevensey Levels. Two guts are shown on OSX 123 at TQ 6106 with others at NGR NZ3166 and NZ2984.

guter (ME) See gutter.

guttar (ME) See aquaductum and gutter.

gutter (E) From the F *gouttière*. Loudon (1243) gives: "a furrow-channel or drain" and Jackson (190) states that a gutter is: "a narrow (natural) water-course, generally flowing into a brook" and gives, for example Hope Gutter, which is mapped at NGR SO5078. Halliwell (425a) gives: "A small stream of water deep and narrow. Yorksh" and WW give the ME forms: 'guttur, guttar and gotyr", (1:587.30; 732.21; 799.44) respectively. Wycliffe (779a, from Psalms 41:9) gives: "*gooteris*" and (734a, from Habakkuk 3:10): "*guter*". Callis (100) states that: "I take it, a gutter is the dim. of sewer: and the difference between them is, that a sewer is a common public stream, and a gutter is a straight private running water; and the use of a sewer is common, and of a gutter peculiar ...." A gutter is shown on OSX 307at NZ0145. See gurgis; guttorium and the MED under goter.

gutterdick (E) Parish (52) gives: "A small drain."

gutter-drain (E) An agricultural drainage or irrigation term. Loudon (1:401:2642) and Rees (34) both mention this term which is used in grasslands.

guttorium (L) WW (1:587.30) give this and (732.21), "*guttatorium*" from two of the vocabularies, as ME: "guttur and guttar" respectively and CTM (254a) gives the variant forms: "guttera, gutteria, guttural and gutturus."

guttur (ME) See gutter.

gutur (ME) See aquagium and gutter.

guuer (Co. & Wel.) Williams (1865:191b) gives: "a brook ... an old form of gover." This is also the oldest form of gofer found in Wel. The 12th Century LL has many entries for it – four are given on p 207, see the index on p 404 and gofer.

guver (Co.) Jago's entry (1887:133a) for this gover variant comes from Borlase (435a) who gives: "a brook".

guy? (Co.) This is one of six cognate forms given by various authorities meaning 'water or river' – the others are gwy, gy, uy, vy and wy. However, most of these forms, with the addition of guai; gui; guye and gvy, given in the (LL), are all variants of the Wel. RN Wye. Borlase (461c) lists guy and uy as found in the termination of names. Baxter's entry (265) has much the same and Jago (1887:176a) follows suit. Pryce lists gwy, gy, vy and wy, which are all confirmed by Williams or Jago in their dictionaries; not surprising, as they both follow Pryce. Bannister (1871) in his *Glossary of Cornish Names* mentions gwy often – again, no doubt taken from Pryce's findings as does Charnock (1859 & 1870). Owen Pughe's Welsh dictionary (1873:207b) gives wy as a suffix to a great number of RN's, in much the same way as Pryce and his followers have done for Cornish names, but, today, all of their definitions have been superseded by later scholarship. Possibly the best example of any of these 'terminations' that can be seen on a map, is the Wel. example of wy, in Bachawy – OSX 188 at SO1746. Bannister (1916:12) mentions another stating that: "There is, however, in Montgomeryshire a brook called Bacho Brook which in the Brut (under year 1111) is Bachwy." In Wel. gwy simply means 'water' and not 'river', however, Pughe & Pryse (1:643a) give: "a stream." On The RN Wye – Gwy in Wel., the Derbyshire Wye and the Surrey Wey, all connected, with some differing opinions by PN scholars, see the DPNW (498); PNDb (19); PNSr (7) and Watts (668a & 706a). Historically, all of the forms quoted herein, with the exception of the LL entries, were, no doubt, given in good faith at the time, however, since then, none have found their way into a current Cornish dictionary and, therefore, are all questionable as qualifiers, but are given for what they may be worth.

guzzen (E) See guzzle.

guzzle? (E) Definitions differ for this and its variants. Wright (1880:525b, 540b & 541a) has five different forms and gives: "gozell a ditch"; "gushill, a gutter"; "gustrill a dirty gutter" and "guzzle and guzzen a drain or ditch; a small stream". Sternberg (45) in addition to guzzle gives: "gudjil, a drain." Baker (1854:1:299) offers: gushill, "a running gutter"; Halliwell (425a): "guzzle, a drain or ditch. South. Sometimes, a small stream. Called also a guzzen" and Dartnell (72) states: "a filthy drain." It is quite obvious that these terms lean more towards sewer definitions than generic watercourses but the fact that 'small stream' is mentioned gives guzzle, or one of its variants, some currency in the questionable category.

gwaz (Bret.) See goeth.

gwely afon (Wel.) Pughe & Pryse (1:655a, under dyfrle) give this, as a compound meaning: "the channel or bed of a river."

gwerdhour (Co.) See guerthour.

gwote (E) Prevost (84b) gives: "A gutter through a hedge, not covered in but stopped up with thorns, &c.; if covered in it is called a cundeth." See gote.

gwter (Wel.) The Wel. form of the E gutter, which can be seen on OSX 262 at SH2990.

gwy (Co. & Wel.) See guy.

gwyffos (Wel.) Evans (1:54a & 2:1063b) gives "aqueduct" and "watercourse" respectively.

gwyle (E) See goyle.

gwyth (Co. & Wel.) Defined by Williams (1865:204b) as "a vein". Bond and Nance (1978:248b) give "stream". For the Wel., GPC (1790c), in addition to vein, has a number of watercourse definitions which confirm the Pughe (2:384b) entry given under "cwtter". See goeth and CPNE (122).

gwythen (Co. & Wel.) This is a dim. of gwyth. Under goth, Nance (1978:249a) gives: "gwythen and gothen, stream". In Wel., the GPC (1791b) lists gwythen alongside gwythïen, defining it as: "small natural channel within the earth through which water trickles", a definition then, which suggests the dims. 'brooklet, streamlet' for gwythen and, probably, gothen too. For the -en terminal compare goveren and goferen.

gwythïen (Wel.) See gwythen.

gwyth melin (Wel.) This compound is given by Evans (2:341b, under mill-leat and millrace). See gwyth.

gwythred (Wel.) This dim. term is listed in the GPC (1792b) defined as: "channel, ditch, canal, brook", dating from 1803.

gwythreden (Wel.) Another dim. of gwyth, variously given as: 'brook, watercourse, rivulet' by most authorities. See gwythred.

gy (Co.) See guy.

gye (E) Wright (1880:541a) gives: "A salt water ditch. Som."

gypsey (E) See gipsey.

gypsey race (E) See gipsey.

gypsies (E) See gipsey.

gyte-streám (OE) B-T (496) give: "*A current, flowing stream.*" WW (1:183.8 & 389.8) give: the L definition, "reuma" See wæter-gyte and PNYW (7:197).

H

had-loont-rean (E) Wright (1880:543b) gives: "A gutter or division between headlands and others. North." Nodal (150) and Grose (77) the same.

hag(g) (E & Sc.) A watercourse through marshy ground in

moorland shown on OSOL 30 at NY8407 and NY8606. The Sc. form is hag, shown on OSX 334 at NS6432.

hail (Co. & E) From IE **sal*, 'salt, water, stream', in L *sal* means 'salt'. Salinae, the Roman name of Droitwich, means 'salt works'. Pokorny (3:878-9) discusses the sal- stem and it is commonly applied as a prefix to many RN's in Europe. In the OPr.EV (170A), *salus* is given as: "brook, (rill)" and in Su., *saluze*, equates with 'rivulet'. There are many forms of hail – this one heads the list – the others are haile, hayle, heil, hel(l) and heyle. Pryce gives "hail(e)" and "hayle", discussed by Norris (489-90) and Bolase lists "hail, heil, hel(l)" and "heyle", repeated in Jago (1887:133a). The various hail forms, taken from 18th and 19th C dictionaries, in all instances, mean 'river' or 'a salt water river'. Today, contemporary works, CPNE (127) amongst them, define the term 'estuary'. In Hunts. (PNBdHu 7-8) is a lost R Hail, early forms of which correspond with the Cor. In Warks. there is another R Heile (Huile in Lewis 1845:1:536) mentioned in Drayton's *Polyolbion* (2:154) "*And as she thence along to Stratford on doth strain, Receiveth little Heile the next into her train.*" This Heile is probably the R Dean which flows through Wellesbourne, according to Drayton's order. Duignan (1912), ERN or PNWa do not appear to mention this river, however, the Ekwall theory, outlined in ERN (188) on the Gloucs. Hail, is discussed in PNGl (2:15). See Ekwall ERN (188 & 192); Watts (290a) and Nicolaisen (243-4) for a detailed and persuasive argument regarding pre-Celtic origin for the RN Hail and its continental counterparts.

haile (Co.) See hail.

hals (OE & ON) See hause.

ham? (E) A watercourse named Black Ham is shown at NGR TL2191; the only evidence for this term, other than a photograph on Geograph, where it is supplemented with "drain".

hamps? (Wel?) In the DBO, under *Origins of Place Names*, Hamps is given as a Celtic term for a river that becomes: "Dry stream in Summer", much the same as winterbournes do. There is a R

Hamps in Staffs., shown at NGR SK0457, the earliest recorded form of which is Hanespe. The name is identical with the Wel. Nant Hafhesp shown at NGR SH9337 which is mentioned by Drayton (2:49) in line (111) from the tenth song – *"Her handmaids Manian hath, and Hespin, her to bring".* The Wel. PN Aberhafesp, shown at SO0692, has a watercourse named Aberhafesp Brook. Morgan (224) gives: "Aberhavesp. – The place is situate at the confluence of the R's Havesp and Severn; hence the name. Havesp signifies a R whose channel is dry in the summer. English name— Hespmouth." The Wel. words haf and hesb, mean 'summer' and 'dry' respectively. The Afon Alun is another Wel. R which tends to dry up in certain places especially at Hesp Alyn (SJ1865) as does the Afon Hepste at SN9511. Hamps in E, only exists as a RN, so, if hamps *is* a true generic, Celt. or otherwise, it is surprising that there are very few references to it. See ERN (190) and DPNW (9, 15 & 194).

hapa (Hitt.) See avon.

hapi (Luw.) See avon.

hause (E) From the OE and ON *hals.* Not normally defined as a watercourse but is marked as such on OSOL 7 at NY3107. The OED1 (5:45a), under halse, hals gives: "A narrow neck of land or channel of water". The EDD (3:35a) defines it: "A defile, a narrow passage between mountains; a narrow connecting ridge." The term has obviously developed a 'stream' sense, as other terms for ridges, cracks, crevices and the like have; such as hearne. See EPNE (1:226).

haven (E) This is a generic watercourse term, but is not given in dictionaries as such. The normal definition is 'a safe anchorage for ships'. A concentrated set of 'havens' can be found on the Pevensey Levels, west of Bexhill in East Sx., see OSX 124 at TQ6607; 261 at TF3541 for another and 293 at TA1728 for Hedon Haven in Yorks., which PNYE (6) refers to.

hayle (Co.) See hail.

hea (OE) A variant of ea given in BCS 1005, from the Cotton Claudius C. ix, 199v: copy of bounds, s. xii manuscript. See ea.

head main (E) An agricultural irrigation term given by Dickson

(1805-7:2:430) and Loudon (726:4411) who states: "Head main is a term used to signify a ditch drawn from the river, rivulet, & c. to convey the water out of its usual current to water the lands laid out for that purpose, through the means of lesser mains and trenches."

head-race (E & Ir.) The watercourse which brings water to the mill wheel – as opposed to the tail-race – which takes it away. See Headrace Canal, Co. Clare, Ireland, built much the same as any other headrace for flour mills but this one was built to divert the waters of the R Shannon to a turbine at Ardnacrusha and is, therefore, much larger.

hearne (E) From the OE *hyrne*, 'recess, curving valley'; not normally defined as a watercourse, but is marked as such on OSOL 31 at NY9017. A term which has obviously developed a 'stream' sense the same as hause, See EPNE (1:276).

heat (E) An alternative name for a gurt or trench – The OED1 (4:516c) under gurt, gives: "A heat, gurt or trench."

hedge-trough (E) See hedge-trow.

hedge-trow (E) Chope (99) gives: "the ditch or drain at the side of a hedge." and adds; "Never ditch-trow." This conflicts with Elworthy's definition (1886:332) who states: "The ditch or drain at the side of a hedge, called more often a ditch-trow in this latter case the trow, i. e. trough, is of course redundant." Wright (1880:561b) gives: "hedge-trough, "A ditch. Devon." It is obvious that the meanings differ in Dev. and Som., but both are valid.

heil (Co.) See hail.

hel(l) (Co.) See hail.

hell-beck (E) Axon (81) gives: "Hell-Becks, little Brooks in Richmondshire, which are so called from their Ghastliness and Depth." Richmondshire is a local government district of Yorks.

helleck (E) This is a rare term. Wright (1880:563a) gives: "rivulet."

henting (E) Ellis (3:52) gives this term as an alternative for water-thorough.

herbery (Sc.) Warrack (259b) gives this rare term as: "a stream."

heyle (Co.) See hail.

hill-burn (Sc.) Warrack (262b) gives this rare term as: "a mountain stream."

hlimme (OE) B-T (544) give: "*a torrent.*" See lumb.

hlinc-gelád (OE) B-T (Supp., 552) give: "*A watercourse on a slope (?).*"

hlípe-burna (OE) B-T (Supp., 552) give: "*A brook with a fall in it*"

hlynn (OE) See lumb.

hoewal (Wel.) This and its variant hoewel is given by Spurrels (242a) as: "stream, channel", confirmed by the GPC who use the forms hoywal and hoywel.

hoewel (Wel.) See hoewal.

hohle (E) Peacock (1889:274) gives: "A wooden tunnel under a bank or road for the conveyance of water." He also gives the forms owle and howl from various quotations therein. In another work by Peacock, *Ralf Skirlaugh* (1870:2:87) he gives: "howle." See The EDD (3:199B & 260b).

hole (E) From OE *holr*, "a hollow, deep valley." Not normally defined as a watercourse in dictionaries but has developed into one much the same as hause and hearne have – at least three can be found on the modern map, an estuarial one on OSX315 at NY3164 and two inland examples are given on OSOL 42 at NY5996 and 7498. See EPNE (257-8).

holl? (E) The EDD (3:210a) gives: "A wide ditch of water." Forby (1:91) under deke-holl, states: "Or *holl* may be used alone in the same sense." That is: "as a hollow or dry ditch."

hollow (E) Not normally defined as a watercourse but is marked as such on OSOL 1 at SK0992.

hollow drain (E) An agricultural drainage term. Young (1799:241) mentions: "hollow drains filled with stone". Rees (34) uses the term: "Hollow Earth-Drain."

hollow earth-drain (E) See earth-drain and hollow drain.

hollow furrow drain (E) An agricultural drainage term. Loudon (709:4294) mentions that: "The hollow furrow drain is only used in sheep-pastures. Wherever the water is apt to stagnate."

holm(e) (E) Robinson (1855:84) defines this term as: "a brook or beck." Atkinson (1868:268) points out that Robinson mistakenly took holm to mean brook, on the other hand, a 'holme' is shown at TG4911, as a watercourse forming part of the boundary of West Caister Civil Parish in Norfolk!

holr (OE) See hole.

hope (E & Sc.) Leland (1:77) refers to: "hopes or bekkes". Brockett (99) gives: "Hope, a small brook, or the valley through which a brook may run; as Stanhope, Bollihope, &c. Durham." Quite a few are shown on maps – three can be seen around NGR NT7712. A Sc. hope is shown on OSX 329 at NS9920 and the variant houp on OSX 336 at NT0129.

houp (Sc.) See hope.

howe (Sc.) Possibly a variant of the E eau, not normally defined as a Sc. watercourse but is shown as such on OSX 427 at NK0636.

howl(l) (E) See hohle.

howle (E) See hohle.

hoywal (Wel.) See hoewal.

hoywel (Wel.) See hoewal.

hrin (E) See wring.

huche (Sc) See sheugh.

hullett (E) According to Rye (108) a hullett is: "a brook with woody banks."

hulve (E) Gepp (18) gives: "a water channel under a gateway." See wolf.

humble-bummel (Sc?) Wilson (1882:236A) lists two places relating to this term as: "sonorous cataract on May and Lednock Rivulets in Perthshire", therefore, a term which favours a waterfall definition rather than a watercourse one.

hush (E) A stream used to wash away earth and stones from minerals, found particularly in use at lead mines. There is a Providence Hush in the Yorkshire Dales next to a lead mine and two others are shown on OSX 298 at SE0668 and OSOL 31 at NY7129. See watersike and PNWe (2:111 & 265).

hyrne (OE) See hearne.

hyttynt dur (Wel.) See cerrynt.

I

ick (Co.) Pryce gives: "ick, ek, a frequent final termination, and sometimes means *a creek, rivulet, or brook.*" Borlase (438a) offers the variants "ik" and "yk". These terminals feature in Bannister's *Glossary of Cornish Names* and as suffixes in goverik and goveryk (dims. of gover – 'brooklet or streamlet') but, independently, are not confirmed as watercourse generics by later scholarship.

igh (SGael.) Dwelly (539b) defines this term as: "burn, small stream with green banks."

ik (Co.) See ick.

imbrocus (L) Martin (260b) gives: "brook" and Black (591b) states: "A brook, gutter, or water-passage."

ings (E) A term given for a number of watercourses by MWDB.

inlay (Sc.) "The channel carrying water to a mill-wheel." given by The DSL (SND).

inlet (E) Peacock (1889:288) gives: "A branch drain used for conveying water from a warping drain to the land to be warped."

inney (Co.) See auney.

innings (E) This term normally defines reclaimed flooded land such as that found in Romney Marsh, but here the term has been applied to the stream which has done the draining, shown on OSX 125 at TR0125.

insouling (E) Peacock (1889:289) gives: "The outfall of a ditch or drain ; sometimes the drain itself; sometimes also a soak-dyke." See the MED under insolling.

intake (Sc.) Warrack (287a) gives: "the part of the body of flowing water taken from the main stream; the place where this water is taken off." A number of intakes are shown at NGR NN292. See The DSL (SND).

irriguum (L) See issiguum.

irriguus (L) See issiguum.

issiguum? (L) WW (1:129.8), give: "stream, *uel* wæto", from *Abbot Alfric's Vocabulary*, related perhaps, to irriguum or irriguus, which has much to do with water.

ithan (SGael.?) Only in Milne (1912:48) as: "stream." The -an terminal suggests a dim. form, but as with so many of Milne's terms, no other authority seems to support the fact that 'ithan' exists as a generic term.

K

kahen-ryd (Co.) See chahen rit.

karrag? (Co.) Jago (1887:133a) states: "Karrag, B. [Borlase] This (karrag) is a doubtful word; but Pryce applies the term *carrog* to a brook." There is confusion here – Borlase (439a) has no entry for karrag, but does list karrog as a "brook or river." Karrak is defined as a "rock" and Pryce correctly identifies carrog as "a brook." Therefore, the karrag form must remain a doubtful qualifier. See carrog and karrog.

karrog (Co.) Borlase (439a) and Lhuyd (141b & 284b) both give this as 'a river or brook'. See carrog and karrag.

keechan (Sc.) Gregor (93) and Warrack (300b) both give: "a small rivulet." The common form is caochan, which is SGael., not Sc.

keld (E) From the ON *kelda*, 'spring, well'. Marshall (1796b:31a) gives: "a spring; or perhaps a general name for a river or brook which rises abruptly ..." and Robinson (1876b:68) mentions that the term is: "often used of a brook, or spring." Two keld's are shown on OSOL 10 at SD9798.

kennel (E) See canel.

kerrynt dur (Wel.) See cerrynt.

keynres (Co.) George (1998:75a & 2009:356b) defines this term as: "torrent, brook." Nance (1978:261b) the same. See chahen rit.

khahen-ryd (Co.) See chahen rit.

king-gutter (E) An agricultural drainage term given by Elworthy (1886:402) who states: "The principal drain in draining a field."

korrnant (Wel.) See cornant.
kouer (Bret.) See gouer.
Kreck (G) See creek.
kreek (Fl.) See creek.

L

la(a)nder (E) Wright (626a) gives: "Launder, a gutter, or channel for water." Pegge (109) gives: "Lander, a long wooden trough to convey water to a distance" and Heslop (439) gives: "LANDER, LAANDER, a gutter or channel for water...." Elworthy (1886:419-20) states: "LAUNDER, A trough or shute for conveying water. This is more properly a Devonshire word, where I have heard it used, somewhat beyond this district; it is very common amongst the miners of Devon and Cornwall, according to Mr. Worth."

laca (Co.) Nance (1978:262b) gives: "rill, runlet, small stream." Williams, (1865:227b) under lacca, states: "According to Pryce, it also means a rivulet", but Pryce's entry actually gives: "lakka, *a well, a pit*, or rather *a rivulet*, which we still call a *lake*, and *leak*, or *leate*." In addition, Jago (1887:92a) mentions that: "They still call a leat or mill-stream a lake, at Lostwithiel." See lake; leat(e) and CPNE 141.

lacca (Co.) See laca.
lace (OE) See lacu.
lache (G & ME) See latch.
lacu (OE) B-T (Supp., 599) give: "*A stream, water-course.*" Forsberg (4) discusses lacu in depth and all the BCS charters (788 for instance) give the form *'lace'*. See lake; EPNE (2:8); ERN (234); PN Brk (782); the MED under lake and Middendorf (84, for a full listing). Many other EPNS volumes give lacu too, they are: Bd; C; Db; Gl; Hunts; Nth; O and YW.

*lacuc (OE) A postulated dim. form of *lacu*, 'small stream'. Forsberg (2) discusses lacuc in depth. See Ekblom (110); Goodall (198); PNW (102); PNYW (7:216) and Watts (364a), who questions the former PN references.

lad (SGael.) This and its dim. ladan together with laid are given by Dwelly (562a) as: "watercourse, mill-lead". See lade.

lád (OE). See lade, lead and lode.

lada (L) CTM (268b) gives: "a watercourse." See lade.

ladan (SGael.) See lad.

lade (E & Sc.) From the OE *lád*. The later form of lead; 'a channel for carrying water to a mill'. Kersey gives: "Course of water." Heslop (435) has: "LADE, LODE, an aqueduct or channel which carries the water to a mill." For the Sc., Warrack (316b) gives: "a watercourse leading to a mill", and Jamieson (3:70): "LADE, LEAD, MILL-LADE. The canal or trench which carries the water of a river or pond down to a mill." An E lade is shown on OSX 274 at TF 4445 and laids on OSOL 2 at SD7192; a Sc. one is shown on OSX 367 at NT1484. See lad; lada; lead; ge-lád; lode; and wæter-gelád; EPNE (2:8-9); ERN (234); PNBk (79-80) and, for the postulated OE *lád*, which features in a number of PN's, see EPNE (2:11); PNBrk (3:886); PNDb (3:739) and PNYW (7:217).

*læc(c) (OE) See *lece.

*læce (OE) See latch and *lece.

*lǽd (OE) See lade.

lækr (Ic. & ON) The IED (403b) gives: "Lækr, lœkr, a brook, rivulet." According to PNYN (207), Leake is rooted to this element, as is Leck, Lancs., PNLa (184); Leek, Staffs., Horovitz (2:411) and East and West Leake, Notts., PNNt (252-3). Names assigned with lækr or lœkr are difficult to separate from lec or lece. The DB records for these places all have a Lec-stem and as pointed out by a number of authorities, Ekwall, for one, in his PNLa (184): "... the name is found so often in England ... that it is difficult to believe that the O.N. word is always the source." What we do know is that, the further east we go, the more likely we are to encounter ON elements in PN's. See IED (403) for lækr compounds suffixed -fall; -far; -gil; -rás and -spræna – all watercourse terms; EPNE (2:26) and PNYW (7:221).

lǽt (OE) See leat.

lag (SGael.) The frequency of 'lag' on the modern map suggests a transferred sense to watercourse, just as other terms for hollow, cavity, pit and dell have become. Three examples are shown on maps – OSX 363 at NS1280, OSX 364 at NN3514 and OSX 419 at NJ1920. See sruthlag.

lagu-streám (OE) B-T (616) give: "*Sea, stream, river, water.*"

laid (SGael.) See lad and laid drain.

laid-drain (Sc.) An agricultural drainage term. Warrack (317b) gives: "a drain in which stones were so laid as to give free passage to the water", and Jamieson (3:73b) states: "A drain in which the stones are so laid as to form a regular opening for the water to pass."

laik (OSc.) See lake.

lake (Co., E, Ir., ME & Sc.) From the OE *lacu*, "*stream, watercourse*", (B-T supp 599). The same form is found in ME, recorded in 1350; For the Co. see laca. In E., lake is well attested. Pulman (1871:110) belived that lake was: "peculiar to Devonshire" but, it is actually much more widespread than that – quite a number of counties have PN's which all root back to the OE *lacu* – PNW (372) is just one example. In his *West Somerset Word-Book*, Elworthy (1886:414) states that: "The word is not applied to a large pond or sheet of water, but always to running water" and Dartnell (90) gives: "a small stream of running water" for the Wilts. word. Map evidence comes from Herefordshire where a lake can be seen on OSX 203 at SO4270. The term is also marked on OSIDS 50 at N9538 in Irl. In Sc., lake is not commonly attested as 'running water', but having said that, both OSc. and ModSc. forms can be found on maps! OSX 461 at ND4484 for *laik* and OSX 321 at NX9595 for *lake*. In Gmy., lake is still used of a stream course, see USBG (GFR:vb) and lak(e) in the MED for a full digest of ME forms.

lakka (Co.) See laca.

land-ditch (E) An agricultural drainage term given by Vancouver (1795:203). See under-ditch.

land-drain (E) An agricultural drainage term given by Young

(1797:155). At least three are shown on maps at NGR SE9816; TA0101 and TF4186.

lander (E) See launder.

land-waters? (E) Holinshed (1:185) mentions this term which would appear to relate more to flooding rather than as a generic term.

lane (Sc.) The SND give: "A swampy piece of ground, a marshy meadow. b. A (slow-moving) stream draining such ground." Chalmers (46) states that: "Loin in the Gaelic signifies a *rivulet*, whence several small streams in Galloway are termed lane, which is a merely modern corruption of the Gaelic word loin." A lane is shown on OSX 330 at NT0011 and the variant loan on OSX 317 at NX2079.

larch-tube drain (E) One of the many agricultural drainage terms given by Stephens (1848:136-7).

latch (E & Sc.) This term comes from the OE **læce*, ME lache and is inextricably linked to letch. Dickinson (194b) gives: "An occasional watercourse." The DSL (SND) have latch and (DOST) gives: "leche". The term is mentioned in RGSS (6:418b), in a document dated 1601. There does not appear to be a mapped example in Eng., but there is a Sc. one shown on OSX 346 at NT8467. In Gmy., lache is still used of a stream. See *lece; letch; PNC (303, 334); EPNE (2:10), which has full lists of associated PN's; PNO (2:456)); the MED, under lech(e) and USBG (GFR:vb).

latex (L) WW (1:29.27 & 435.22) give: "burne" and "burna" respectively; elsewhere (326.3): "burna oððe broc"; 'a brook', taken from various vocabularies.

launder (E) See la(a)nder.

lavant (E) A spring which breaks out, usually on the downs, in wet seasons, feeding a brook that is normally dry, similar, but not quite the same, as nailbourne and winterbourne. See The EDD (3:537a); ERN (263-4); Cope (52) and Parish (68).

lead (E & Sc.) From the OE *lád*. The earlier form of lade; The OED1 (6:139b) gives: "An artificial watercourse, esp. one leading to a mill", and quotes an extract from the Ludlow

Churchwardens' Accounts dated 1541. Kiliani (276b) gives lede, in the same sense and (279b) leyde and water leyde as: "Aquae ductus, aquagium", which shows a Dut., connection. For the Sc., Warrack (324a) gives: "a mill-race, artificial water-course"; one is shown on OSX 367 at NT1097. In Bel., geleed, lede and leed(e) are still used of a canal and drainage ditch, see USBG (Bel:iva). The term is also found in compounds, see by-, mill-, tail- and water-.

leader (E) An agricultural drainage term. Baker (1843:36) gives: "The cost of this method at 4 to 5 yards apart, with the leaders dug and filled with wood and straw, does not exceed 20s. per acre".

leading ditch (E) An agricultural drainage term and alternative name for a carrier or master drain. Young (157), in answer to a question, states: "By a leading ditch, I mean a carrier or master drain, into which all the single drains empty themselves ...."

*leaht (OE) The printed charter in *Liber Monasterii De Hyda*, relating to Leckford in Hamps., gives: *"Leahtford"* twice. BCS (822) gives the same. Grundy, in his comments on '*The Saxon Land Charters of Hampshire with Notes on Place and Field Names* ', Archaeological Journal, 2nd series 33 (1926), 91–253, states: "Probably, though not certainly, *Leac-Ford,* 'Ford of the Leeks.' ... But Dr. Bradley suggests *Leaht-Ford,* 'Ford of the Irrigation Channel'." The R Test around Leckford has many channels and drains along its course so we may conclude here that 'ford of the irrigation channel' is the right derivation. See EPNE (2:22).

leak (Co. & E) For the Co. see laca. In E., one of the few definitions given, comes from Wright (1880:628b) who gives: "A gutter. Durham."

leam (E) Davies (82) gives: "a drain or watercourse." Smiles (1861:1:67) mentions it too, and two leams can be seen on OSX 235 at TF4525. On the RN Leam and other related names see BCS (978, S623) *"limenan"*, hence the connection with Leem-; Lem- and Lym- names; ERN (243-6); PNRB (385-6) and PNYN (227).

leart (E) See leat.

leat (Co. & E) This comes from the OE *lǽt*, found in the compound wæter-gelǽt, listed by WW (1:211.13) and B-T (1161a), the Co. term is, no doubt, borrowed from E. Thomas (100) defines leat as: "a small river, a stream." Courtney & Couch (34a) give: "Leat, a gutter; a narrow artificial water; a mill-stream." Jago (1887:92a) states: "artificial water channel" and Pryce lists "leate" under "Lakka". In E, the term is quite widespread; there are many references – one such, from Elworthy (1886:424) gives: "The water-course leading to a mill." Pulman (1871:111) has the unusual spelling "leart" – there are map references too – leat is shown on OSOL 5 at NY3516 and leet on OSX 132 at SU4722. The old form of Longleat, in Wiltshire, "Langelete", dates back to 1257 and can be seen in AD (3:442:D319) and PNW (169). See laca.

leate (Co.) See leat.

*lec(c) (OE) See *lece and letch.

*lece (OE) The OE elements **læc(c)*; **læce *lec(c)* and **lece* are difficult to separate – they have been assigned to many PN's and words such as latch and letch, and it is generally agreed that they correspond with the ON *leka*. Forsberg (72) mentions the Lece Brook from a charter in Dugdale's *Monasticum Anglicanum* – probably the most relevant. One of the best listings for PN's with the OE **læc(c)*; **læce *lec(c)* and **lece* elements can be found in PNCh. (5:1:ii) and the DB lists over thirty PN forms beginning with the Lec- stem. See PNBrk (886); PNDb (3:687, 738); EPNE (2:22); ERN (246) and PNGl (4:147).

lecha (E See letch.

leche (OSc. & ME) See letch.

lede (Dut. & Fle.) See lead.

leed(e) (Fle.) See lead.

leet (E) See leat.

lèig (SGael.) A rivulet running through swampy ground – shallow stream – a low valley, a depression – a pit or ditch of water; just some of the definitions given for this word and the following variant forms. Dwelly gives "leoig" (583b) "log" and "loig" (595a). Jamieson (3:130a) has "leog" as does Edmonston (65) and the SND uses the form "lyog" and gives other variants. Angus, in his *Shetland Glossary* (86), lists "ljoag" as does the SND. Jakobsen (520a) and (534a) lists the SNn. forms "ljog, log and løg". The various forms of this widespread element have obviously come down to us as dialectal developments, supplemented, and probably influenced, by some of the Norn forms, found and recorded in the Shetland Islands, particularly, by Jakobsen. The only form recorded on the modern map is leig, which can be seen on OSX 458 at NB2338 and OSX 460 at NB5346.

leka (ON) See *lece.

leme (ME) For watercourse confirmation see EPNE (2:23); PNGl (4:151) and PNYW (7:220).

leog (S & O) See log.

leog (Sc., SGael & S & O) See lèig.

leoig (SGael.) See lèig.

letch (E & Sc.) From the OE *lecc*, ME *leche*. The AD (2:446) give: "le Siche called Leche"; Morris (1857:42) has: "lecha", and Wright (633a): "A wet ditch or gutter. North." Heslop (448) states: "LETCH, a long, narrow swamp in which water moves slowly among rushes and grass." Sc. defintions are given in The DSL (SND & DOST) and two examples are shown on OSX 332 at NU1209. See latch; *lece and EPNE (2:10).

leth (Sc.) Jamieson (3:134a) gives: "A channel or small run of water."

Ley (G) See lead.

leyde (Dut.) See lead.

***licc** (ME) Forsberg (102-3) discusses this element under Beoforlic (Beverley, Yorks.). Beoferlic is the form given in the ASC (1:73b). BCS (644 & 645) gives: *Beuerley* and *Beverlike* respectively. The DB form is Bevreli. Although a stream definition has been suggested as the second element of Beverley, opinions, amongst PN scholars vary. See streamlic;

EPNE (2:24); PNYW (7:220) and, for a discussion on how OE
*lecc became *licc*, PNYE (194).

lin(n) (E? & Sc.) Lucy Toulmin Smith's Glossary, in Leland
(5:xxvii) gives: "Lin, a linn, waterfall or torrent, but Leland here
uses it for a small stream in low land, i, 95." In Sc., lynn is
sometimes used as a a variant and all the mapped examples are
in Scot. too – OSX 330 at NT2814 and OSX 343 NS9157,
show one each, with another four showing at NGR NY0089.

linne (SGael.) This is given by Dwelly (590a) as: "mill dam,
channel, cataract, waterfall."

little-wawder (E) Obviously classified as a tributary by Pulman
(1871:100) who, under gurt-wawder, states: "A river proper, as
distinguished from its tributaries, which are all Little-wawders."
See gurt and gurt-wawder.

ljoag (Shet.) See lèig.

ljoag (SNn.) See log.

ljog (SNn.) See lèig.

ljog (SNn.) See log.

llednant (Wel.) Spurrell's (257b) list this compound as:
"tributary", also given in the GPC (2431c).

lli (Wel.) Evans (2:886a, under stream) gives: "lli, lliant and llif'.

lliant (Wel.) See lli.

llif (Wel.) Lhuyd (284b) uses the form: "lliv". This term is
normally defined as: 'torrent, stream, flood'. Map references
exist too, but only in the form llyf-, compounded with nant. A
llyfnant can be seen on OSOL 23 at SN7497 and another west
of Newtown at NGR SO0689. See lli and lyf.

llif ddwfr (Wel.) Pughe & Pryse (2:285b) give: "a stream of water,
a torrent" and the variant llif ddwr, for which Salesbury gives:
"flowyng water".

llif ddwr (Wel.) See llif ddwfr.

llif dwfr (Wel.) Evans (2:886a, under stream) gives this
compound. See llif ddwfr and llif ddwr.

lliv (Wel.) See llif.

llyfnant (Wel.) See llif.

llyr (Wel.) Evans (2:1063b) and Pughe & Pryse (2:305b) give this term as "watercourse". The nearest mapped form is in lleiriog, the name of a stream, which can be seen on OSX 239 at SJ1525.

llyry (Wel.) Evans (2:886a, under stream) gives this form. See llyr.

load (E) This form is shown on the Ely to Downham Market map, dated 1645, by J Blaeu. See lode.

loan (Sc.) See lane.

lock-furrow (E) Wright (643b) gives: "A furrow ploughed across the balks to let off the water. South."

lode (E & Sc.) From the OE *lád*. Wright (642a) gives: "Load" and "Lode", A ditch for draining the water from fens" and Forby (2:199) states: "An artificial water-course"; Britten (104) has: "Load. A lode, a water-course." The OED (6:367b) state that: "the words load and lode are etymologically identical." A lode is shown on OSX 226 at TL5679 and a Lode village and course at NGR TL5262. See ge-lád; lade and PNC (131 & 335) where all the Cambs. lodes are discussed under ge-lád.

lodge (E) Not normally defined as a watercourse but is shown as such on OSOL 19 at NY7929.

lœkr (Ic. & ON) See lækr.

log (SGael & SNn.) For the SGael., see lèig. Jacobsen (1928:520a) gives: "log, ljog, a small, quietly-running stream of water", and, (1928:534a): "løg, a brook." He also lists (1897:87): "Ljoag is a patch of green, through which a streamlet runs." It comes from the ON *loekr*. Edmonston (65) gives the S & O form: "Leog, a rivulet running through low swampy ground." See lèig.

løg (SNn.) See lèig and log.

loig (SGael.) See lèig.

loin (Sc.) See lane.

lòin (SGael.) Dwelly (595b) gives: "little stream, rivulet", and (598a) defines lón as: "small brook, especially with marshy banks." Charnock (1859:92), under Dunfermline, gives linne or loin as a termination for the PN but most authorities question the second element. A loin can be seen on OSX 420 at NJ2529. See lane.

lón (SGael.) See lòin.

lone (Sc.) See lane.

loop (Fle., MDut. & Sc.) Jamieson (3:171a) gives: "The channel of any running water, that is left dry, when the water has changed its course.... This term is of very ancient and general use as denoting the course of a stream", and adds that it comes from the Ice. "hlaupa". Warrack (337b) gives: "The channel of a stream left dry by the water changing its course." Kiliani (292a & 648b) gives: "loop der riuieren. Alueus, fluuij, fossa per quam labitur flumen." and "waeter-loop. Aquagium, aquae ductus", respectively, presenting us with a MDut. connection. In Bel., the Fle. loop and waterloop are still used, defined as: "drainage ditch, stream." See USBG (Bel.:iva & ivb).

lossan (SGael.?) Only in Milne (1912:24) as: "small river." The – an terminal suggests a dim. form, but, as with ithan, no other authority seems to support it as a generic term.

low (E) Heslop (458) states that "The tidal stream at Goswick is called Goswick Low", but it is not shown on the OS map; only South Low is shown here which runs north to south passing through the causeway to Holy Island. There are 6 'lows' marked on OS OSX maps: Engine Low NU0543; North Low NU0445; South Low NU0643; Black Low NU0841; Ross Low Nu1336 and The Low NU0240. Heslop's word book does not give low as a watercourse generic but it is quite obvious, from map evidence that 'low', in Nrth., is used of a watercourse in much the same way as creek is used in Norf. See ERN (264-5).

low-shot (E) See over-shot.

lub (SGael.) Dwelly (604a) does not give a watercourse definition for this term only for the compound lùb-shruth (604b). He defines lùib (607a) as: "angular turning, winding or bend, as of a stream." However, both of the simplex terms exist as watercourse generics and can be seen on OSX 459 at NB3027 and OSX 460 at NB4352 respectively.

lùb-shruth (SGael.) See lub.

lùib (SGael.) See lub.

lum (E) See Lumb.

lumb (E) Addy (141) gives: "LUM, a narrow valley containing a stream of water." – discussed in depth (xxviii-xxix), where he connects Lumb with Lim: "... In this moss, or moor, a stream rises which on modern maps is called Limb (properly Lim) Brook ... Lim Brook is probably quite a modern name. The stream itself was formerly called a *lim*, or torrent, and the word is still found in the neighbourhood of Sheffield as *lumb* or *lum*. In Anglo-Saxon the word is found as hlimme." This Limb Brook is shown at NGR at SK3083; PNYE (131) mention it also, but here, it is connected with *hlynn* not *hlimme*, (B-T, hlimme, hlynn, *'a torrent'*). The lim element is all to do with *lime*, so, on the complexities of whole issue of OE *hlimme* 'stream' see EPNE (252), and OE *hlynn* 'torrent, EPNE (254) and for Lymm, PNCh (1:2-60) which mentions ERN (243-6), which, in turn, connects the Lem- and Lym- stems with 'elm'. A connection between the Lim(b) and Lumb names would be very welcome indeed, but, in view of the foregoing, the impossibility is upheld at this time pending further research. There are numerous watercourse and PN's with the lum(b) element – A Lumb is shown on OSX 268 at SK3376; Hallas Lumbs, near Cullingworth, is shown on OSOL 21 at SE0737 (1978 Edn.) but is now marked in black on latest maps, and, a tautological Lumb Dike is shown at NGR SE1513. A Lumb Brook, in Lancs., can be seen at SJ6285 and High Lumb Brook, a tributary of the R Darwen, adds yet another. Lumb Brook in Derbs. is shown at SK3347 and Cheesden Lumb, now marked Cheesden Brook is shown at SD8316. Next up is Lumb Hole Brook shown at SD9710 and another Lum Brook is shown at SE1226. In Ches., Lumb names are common too – a Lumb Brook, a tributary of the R Dean, is shown at SJ8881. All these lumbs suggest that lum(b); a 'valley'; 'a valley with a stream' or 'a pool in a river' was a term transferred to the stream at some time in the past, much the same as wham. However, the first mentioned above is sufficient proof that the case for lumb, as a watercourse generic, is confirmed beyond all doubt. See The EDD (3:689b) under lum; EPNE (27) and, on Lum or Lumb as an element in PN's, of which there are many; PNCh (1:30) and (5:2:96) which lists over fifteen Lum; Lumb and Lumm names; PNCu (183 & 478); PNDb (1:33); PNLa. (59, 62, 64, 74, and 93); Watts (386); PNYW (3:148, 7:205 for hlynn) and (8:117) for a list of over thirty Lum(b) PN's.

lyeur (A-N) Kelham (142a) gives this rarity as: "a brook."

lyf (Co?) Nance (1978:179a) gives, under torrent: "fros ... lyf ... keynres", and, under lyf (266b), gives: "flood, deluge." The first entry suggests that lyf equates with fros and keynres, both watercourses; the second, not so, therefore, lyf, as a watercourse term in its own right, might be questionable even though the term exists in Wel. See lli; llif and the llif-compounds.

lymph (E) A poetical stream term given by Darwin (42) *"Call from her crystal cave the naiad-nymph, Who hides her fine form in the passing Lymph."*

lynn (Sc.) See lin(n).

lyog (Sc.) See lèig.

M

main (E) An agricultural drainage term. Loudon (728:4433) gives: "mains" – Wikipedia, under Water-meadow: "main", as does FD, however, *main* is more commonly applied as a prefix in compounds and not well evidenced as an independent generic drainage term. See all the main- terms.

main carriage (E) See main water-carriage.

main-carrier (E) An irrigation term given by Dickson (1805-7:2:437): "... water from the main carriers ...."

main ditch (E) An agricultural drainage term given by Loudon (1146-7) "On the Lilleshall estate of Lord Stafford ... in 1816 and 1817 there has been executed ... 46,000 yards of main ditches made or deepened ...." See all the main- terms.

main drain (E) A very common agricultural drainage term mentioned by Loudon often. There are many mapped instances – a typical example is shown at NGR TF5375.

maine trench (E)See trench-royal.

main stell (E) See stell.

main water-carriage (E) An agricultural drainage term mentioned by Smith (1851:141) in his paper on catch-water meadows and reported in the BFM (20:449) "The arrangement of the "main water-carriages" depends solely upon the formation of the land and supply of water ... These "main carriages" are formed 3 feet wide and 6 inches deep on the lower side."

maister-drain (E) See master drain.

master drain (E) An agricultural drainage term. Young (157), in answer to a question, states: "By a leading ditch, I mean a carrier or master drain, into which all the single drains empty themselves ...". Carr (307) gives: "MAISTER-DRAIN, a principal drain."

master feeder (E) An agricultural irrigation term. Wright (1790:19) gives: "The bottom of the first work, or master-feeder, ought to be as deep as the bottom of the river."

master furrow (E) An agricultural drainage term. The OED1 (6:214c) gives, in a quotation taken from Walter Blith's *English Improver*, published in 1649: "A good Drayne or *Master Furrow*".

master work (E) An agricultural irrigation term. Wright (1790:47) gives: "the master work which waters the highest and most distant part of the land."

meadow drain (E) An agricultural drainage term; Dartnell (23) lists it as a definition of carriage. Marshall (1787:91) gives: "Nothing is more common than to hear of stock being smothered in the meadow-drains ...." and (391) "WATER-WORKERS. Makers of meadow-drains." This term is also mentioned in Fenland N&Q (351) "The climax of this frightful outrage is, that the sister, through despondency from the loss she had suffered, drowned herself on Monday evening last, in the North Meadow Drain ...."

meatus (L) A 'course'. See waterway.

meoir (SGael.) Dwelly (618b) gives meoir and meur as: "branch of a river". Both terms are both well represented on maps –

meoir on OSX 404 at NJ2006 and NH4298, and meur on OSX 472 at NH8440 with a cluster of 5 shown on OSX 440 at NH4896.

meur (SGael.) See meoir.

milestreame (OE) See mill-stream.

milicentum (L) CHR (42) gives: "a millstream."

mill brook (E) The first of many mill- compounds. First recorded in OE, B-T (703) give: "mylen-bróc, *A mill-brook*." Longfellow (272b) gives: "The mill-brook rushed from the rocky height." A mapped example is shown at NGR SU5487.

mill-burn (E & Sc.) From the OE *mylen* and *burna*, B-T (644) give: "mylen-burna" and BCS 695: *"mylenburnan."* OE had other compounds too – B-T (181) lists: "cwyrn-burne, *a mill stream*" and (Supp., 138): "cweorn-burna, *A mill-stream*." The cwyrne and cweorn prefixes are common OE words for a mill sometimes used of a handmill or quern, hence the spellings. The Sc. is better attested – Warrack (358b) gives: "a stream driving a mill" and there is also an entry in The DSL (SND) plus a mapped example shown on OSX 451 at ND2959.

mill-cloose (Sc.) See mill-trowse.

mill-course (E) Stephens (1848:96) gives: "... a mill-course or rivulet ...."

mill cut (E) Cornish (100) gives: "Wherever there is a water-mill, a mill cut is made to take the water to it. The larger the river, the bigger and deeper the mill cut and dam, unless the mill is built across an arm of the stream itself."

mill-eat (E) A corrupt form of mill-leat. See mill-leat.

mill fleam (E) References are given in the Durham Roll (3:636, 662 and 650), in various forms, and a mapped example is shown at NGR SK2029.

mill goit (E) This term is shown at NGR SK2029. See goit.

mill-gue (Sc.) Mackintosh (267) gives: "During her husband's lifetime they had lived at a farm in the Hillside of Birsay, called the "Mill-gue," or the Mill-burn – a "gue" meaning a deep burn or water course." Jamieson (4:752a) gives: "Way-Goe ... place where a body of water breaks out". See The DSL (SND).

mill-lade (O & ModSc.) Warrack (359a) gives: "a mill-race, or its channel." The DSL (SND) gives the variants -ledd and -lead and The DOST gives the variant prefixes "milne- and mylne-." OSX 367 at NT1484 shows a mapped example.

mill-lead (E & Sc.) The OED1 (6:444c) and The DSL (SND) both list mill-lead.

mill-leat (E) Bailey and Phillips both give: "mill-eat, mill-leat, a trench to convey water to or from a mill." One is shown on OSX 126 at SS2111.

mill-ledd (Sc.) See mill-lade.

mill-pot (Sc.) According to The DSL (SND) a mill –pot, is: "the part of the water-channel below a mill-wheel".

mill-race (E & Sc.) Blakeborough (950) gives: "mill-race, mill-reeace, the cut or channel which leads to the water-wheel, the water running towards the water-wheel." Atkinson (1868:338) gives much the same. For the Sc. Warrack (324a) gives: "an artificial water-course." The DOST also list the term. One is shown on OSX 131 at SU3421 and another on OSX 306 at NZ5723.

mill-reeace (E) See mill-race.

mill-run (E) The only source for this compound is Wikipedia, it is given under mill race.

mill-rundle (E) This compound is shown on OSX 274 at TF4475.

mill-stream (E & Sc.). Cornish (142) gives: "the mill-stream bridged by the main street." BCS 687 (S1208), in a charter dated 931, gives: *"milestreame and mylestreame."* B-T (644, under mylen-stréam) give the same. Two E mill-streams can be seen on OSX 174 at TL3812 and OSX 180 at SP5203 and a Sc. one on OSX 382 at NO7162.

mill-tail (E) Elworthy states: "MILL-TAIL … The stream of water as it runs out from under the water-wheel, after having done its work" and Peacock (1889:352) gives: "the waste water from a water mill." One is shown on OSX 145 at TQ0558. See tail of the mill.

mill-trou (E) From the OE *mylen-troh*, *-trog*, B-T (644 & 703).

WW (1:198.25 give: "*mylentroh*, canalis" and Heslop (478) has: "MILL-TROU, the spout carrying water to a mill wheel." Palsgrave (1852) gives: "Myll troughe or broke."

mill-trowse (Sc.) Warrack (632a) gives: "Trowse, Trows, the conduit carrying water to a mill." Jamieson (278a) states: "MILL-TROWSE, the sluice of a mill-lead, Gall. Mill-Cloose, the same with Mill-trowse. Gall. Encycl.; q. the troughs that conduct the water." See mill-trou and trow.

millwash (E) The OED1 (6:445b), quoting Longstaffe (xvii), gives: "... an old bridge over the millwash ...." See Biblio., under Drake.

milne-lade (Osc.) See mill-lade.

mine (E & Sc.) Apparently a Middlesex dialect term used in draining. Ellis (4:192) gives: "... some of the *Middlesex* Farmers, about *Harrow, Stanmore*, and adjacent parts, who make it their business to get a great deal of sullidge out of the bottom of drains in roads, commons, and other places, which they here call a *mine* ...." In Sc. it is used as a mining drainage term; see The DSL (SND) under mine.

mionshruth (Ir.) It would appear that the only reference to this term comes from O'Reilly (363b) who defines it as: "a rivulet."

mire (E & Sc.) This term is normally defined as 'swampy ground, a boggy place' – in Scot., 'a peat bog'. There is a watercourse in the Yorks. Dales which uses mire as a generic, shown at NGR SD9494 and OSX 451 at ND3165 shows two in Scot. with another showing on OSX341 at NS1451.

mole-drain (E) An agricultural drainage term given by Loudon (710): "It is chiefly useful in pasture-lands".

moss? (E) Not normally defined as a watercourse but is marked as such on OSX 133 at SU7936, the only other reference to this term is to be found in PNDb (1:13) but this one was originally suffixed beck.

mother-drain (E) A main drain – there are many shown on maps – one at NGR SE6300 another at SJ2420 and a Little Mother Drain at SK5997.

mother-dyke (E) Much the same as a mother-drain, one is

shown at NGR SD5161.

muing (Ir.) According to Joyce (1869:3:517) a muing is: "often applied to a narrow stream flowing through a marshy bog." Muinganierin, in County Mayo, is one PN example given – the first element derives from this term. There is another muing shown on OSIDS 23 at F8725 – this one prefixes a stream name – muingnakinkee – other muings are shown at F9921 & G0021 and a River Muing is shown at F9820.

mylen-bróc (OE) See mill brook.

mylen-burna (OE) See mill-burn.

mylenburnan (OE) See mill-burn.

mylen-trog (OE) See mill-trough.

mylen-troh (OE) See mill-trough.

mylestreame (OE) See mill-stream.

myll troughe (E) See mill-trou.

mylne-lade (OSc.) See mill-lade.

N

nailbourn (E) Gower (1893:7)tells us that: "These *bourns* are called in Kent Nail burns." Parish & Shaw (106) quoting Harris give: "nailbourn" and "eylebourn". Basically, nailbourns and eylebourns are intermittent streams which suddenly break out of the earth, run for a while and then disappear. A mapped nailbourn can be seen at NGR TR2048 south of Barham in Kent. See bourn.

nailburn (E) See bourn and nailbourn.

nannau (Wel.) Pughe (1832:2:357a) gives nannau and nonau (365a) is quoted from a line in the Book of Taliesin. However, the Taliesin form is nonneu, in the Skene edition (1868:198) which translates 'streams', a plural form. Morgan (162) gives: "Nannau. – A compound of nant, a brook, and au, a plural termination", this is a PN in Gwyneth, formerly Merioneth, shown at NGR SH7420. See non and nonnen.

nanney (Wel.) See nanny.

nanny (E) In Nrth. there is a Long Nanny, which ERN (355) roots to nant. The name suggests that there was probably a 'Short Nanny' as well. If the name does belong to nant then it must be one of the few, if not *the* only instance in Eng. PNNorDu (147) gives: "Nanny River (Bamburgh), 1245 Pipe Nauny. A Celtic river-name." Morgan (34) states: "Nant ..., Nannau and Nanney are plural forms of it ...." ERN (298) discusses the term, and the Long Nanny, mentioned above is shown on OSX 340 at NU2127. See nant.

nant (Bret. & Wel.) From the PC **nanto*. This term is as common in Wal., as brook is in England. Lhuyd (2c, 141b and 165a) gives: "amnis, rivus and torrens" respectively. Evans, Pughe & Pryse and the GPC all give 'brook, rivulet or stream' definitions, backed up by hundreds of examples on the modern map. In addition, there is a River Nant in NW Scotland which can be seen on OSX 360 at NN0126, however, this *nant*, has a different definition according to CPNS (438). Further, Johnston (189) gives the following: "NEANT, R. (L. Etive). Looks like WEL. nant, a stream, or a ravine; but this is a very un-Brythonic region; ? G. neanntag, nettles." This looks like the same RN with a different spelling but with a more realistic final definition. There are other mentions of *nant* in CPNS (360) which *do* relate to a valley and brook definition and The OGS, under Tranent (6:448B) states: "Its ancient name, Travernant, means 'the hamlet on the vale' — from the Cymric tref, 'a homestead or village,' and nant, a valley." The root of nant is related to the Co. nans, 'valley' and goes back a long way to Bret. and Gaul., *nanto*, 'valley'. For the Bret., Le Gonidec (458a) gives: "Torrent, Courant" and Loth (18) gives: "Nanto, Valle." For the Gaulish, Dottin (85) gives: "vallée: gall. nant «vallée », gaul. -nantus" and (274) "nanto, w valle » [Glossaire de Vienne), gall. nant « vallon». fr. savoyard nant". Nanto is also found in Nantosuelta; the name of a Gaulish Godess. In Fra., nant is still used as a watercourse generic; especially around Nant Borrant, Nantbrun, and Nant De Tamié and many other places too. See nanny and neint.

*nanto (PC) See nant.

náshin paáni (Gyp.) Smart & Crofton (113) give: "A stream, running water" and also list (115) "panái, páni, or paúni, as "water." It is interesting to note here that 'pani', together with 'paniu', "water" (Old Indo-Aryan paniyam), is found in Gujarati and Hindi, showing the antiquity of words found in Gypsy.

navigation (E) An artificial watercourse, sometimes appended to RN's such as the Rive Lee Navigation, shown on OSX 173 at TQ3685 – another example is shown on OSX 130 at SU1627.

navvy (E) A canal or navigation, is the normal designation, but in the South Pennines, on the OSOL 21 map some years ago, probably about 1978, a 'navvy' was shown at SE0032 – this was a natural watercourse unconnected with any form of canal or navigation. It is not shown on later editions of the map, but can be seen in the relative square at NGR SE0080232135 on Geograph.

near (SGael.) Dwelly (687a) gives: "water, river", which would appear to be the only reference to this term.

neint (Wel.) A variant of nant given by Pughe & Pryse (2:357a). The nearest form in the GPC (2570b) is "nentig". See nennig, nentig and nentydd.

nennig (Wel.) Evans (2:711a) lists this dim. term under rivulet. Spurrell's (292a) give: "small brook, streamlet", and the GPC (2570b) give nennig as "a small stream." See nentig.

nentig (Wel.) Evans (under brook, 1:191a) gives this dim. variant as "brooklet." See nennig.

nentydd (Wel.) Pughe & Pryse, under cornant, (354b) and under nant (2:357a) give this dim. as a variant of nentig.

net (SGael.?) Only in Milne who lists net (1912:142) as: "stream", netan, (74), "small burn", neth (184), "stream" and nethan (15), "little stream". The –an terminals suggest a dim. form. There is a River Nethan in Lanarkshire shown at NGR NS8146, discussed in CPNS (210-11), but, even this does not support any of Milne's net- forms or derivations.

netan (SGael.?) See net.

neth (SGael?) See net.

nethan (SGael?) See net.

nick (E & Sc.) Ross (1877:100) gives: "a notch ; a cutting ; a drain. A drain cut by a member of the Bethel family, of Eise, Holderness, went by the name of 'Bethel nick'." A nick is shown on OSOL 5 at NY3618 and another on OSOL 26 at NZ7003. Sc. nicks are shown on OSX 336 at NT1383 and OSX 346 at NT7253.

nill (Gyp.) Another gypsy generic listed under river by Smart & Crofton (183).

non (Co. & Wel.) This term is not well documented. Nance (1978:275a) is one of the few, if not the only one, to give the Co. word rendering it as: "streamlet, brook." For the Wel., Pughe (1832:2:377b) gives: "a stream, a brook." It does not appear in later Co. dictionaries or the GPC. See nannau and nonnen.

nonau (Wel.) See nannau.

nonnen (Co.) The dim. of non – Nance (1978:275a) gives: 'streamlet, brook'.

nonneu (Wel.) See nannau.

nullah (E?) Peacock (1889:377) gives: "a drain (probably obsolete)." Not an E word according to the OED1 (6:254c).

nymph (E) Given by Sylvester in a literary and poetical sense (1:78 line 656) "... Kennet, ... Her Silver Nymphs (almost) directly leading To meet her Mistress (the great Thames) at Reading."

O

o (SNn. & SGael?) Jamieson (Supp:1a) states that: "The terminations au, aw, o, ow, are forms of Gael. abh, water ; as in the Awe in Scot., and the Ow in Ireland." Jakobsen (1897:86) and (1928:2:625a & b) discusses this term in depth, as an old word for a burn in the Shetland Islands, but it looks doubtful as a SGael. term as all o terms in Ireland use the ow- stem and in Scotland there is a lack of evidence to support it as an independent generic term. The PN Thurso uses the o suffix which comes from the ON á, 'river, stream'. The same is found in Laxo, 'salmon or trout burn' discussed by Jakobsen above.

ob (OIr.) Hessens (177b) and Stokes (1887:142 & 256) list ob – oba and obadh are in O'Reilly (389b), all variant forms of ab, aba and abh. O'Brien (342a) gives obha and obhuin. See ab and aba.

oba (OIr.) See ob.

obadh (Ir.) See ob.

obann (Ir.) Only in Hessens (177b) as "fluss, river".

obha (Ir.) See ob.

obhuin (Ir.) See ob

offset (Sc.) The DSL (SND) gives: "A diversion of a stream or conduit forming a mill-race."

oin (Celt.) See abhin.

old english drain (E) Stephens (1889:5:262) gives this agricultural drainage term: "... at one time common in the south-eastern counties of England and elsewhere, partially filled with brushwood, faggots, poles, straw, twisted hop-bines, and even horns, the soil being returned to its original place on top of these."

oltan (Mx.) See alltan.

oozelet (E) Carlyle (7:68) gives: "... and intricate meandering little runlets and oozelets."

open cut (E) An agricultural drainage term. Stephens (1848:6) gives this compound under: "*Draining by Open Ditches*. Mere *surface*-draining is effected by *water-furrows*, and *open cuts* and *ditches*," here, as can be seen, it's just an alternative name for an open ditch.

open ditch (E) An agricultural drainage term. Stephens (1848:6) discusses this under "*Draining by Open Ditches*." and further on p 96. Young (1797:158) mentions it too. See open cut (E).

open drain (E) Stephens (1848:10-11) gives this agricultural drainage term : "Now, open drains upon the surface will be quite sufficient to remove all the water that would remain in a stagnant state in winter, and prove injurious to the roots; ...."

open foor (E) An agricultural drainage term. Heslop (571) gives: "The water channel between "rigs" is thus sometimes called a reen or "open foor."

or (OCelt.) See owr.

*or- (IG). See rithe.

orli (SNn.) Jacobsen (1928:639b) gives: "an opening in the base of a stone wall, through which a burn runs", and "worli"; "wirli" (1066a & b), both in the same sense.

os (SGael.) Dwelly (712b) gives: "mouth or outlet of a river." only, but map evidence suggests a watercourse definition is also possible as two are shown on OSX 458 at NB1624 and NB2036.

oth (SGael?) Only in Milne (1912:259) as: "stream." This looks like a contraction of the SGael., 'othainn', but, as with so many of Milne's terms, no other authority seems to support the fact that 'oth' exists as a generic term.

othain (Ir.) O'Donovan (1860:10-11) gives this rare form as a prefix to the RN Mura; the same is in Meyer (4) and it is also found as othainn in SGael. See below.

othainn (SGael.) McAlpine (193b) states that othainn is used to denote: "the largest kind of rivers, and abhainn a secondary river". Dwelly (713b) gives the same. It may well be that othainn was used as a primary at sometime in the past but, today, abhainn has taken over that role. See othain.

oub (OIr.) Hessens (177b) gives oub as a generic term. Stokes (1903:340, notes) states that oub appears as "aub" in the *Lebor Laignech* – the Book of Leinster. See aub.

our (SGael?) Only in Milne (1912:16) as: "burn" also ouran (26) as: "small stream." The –an terminal suggests a dim. form of our, but, regarding the term 'ouran', it would appear that the only PN to be found in Scotland is the anglicized form of *Sgurr Fhuaran* – Scour Ouran, the name of a mountain.

ouran (SGael?) See our.

oure? (Ir.) Only found as a variant of avon, a ME form given in Spenser's *The Faerie Queen* (235a, Canto X1, xliv, line 5). Joyce

(1911:90) identifies the 'Oure' as Avonbeg, in Wicklow, Avonbeg is *'Abhainn Beag'* in Irish, Avon being the anglisised form is appended here much the same as the variant *'Owen'* is appended to other RN's in Ir. Oure, as a generic variant of avon is doubtful and, as such, remains in the questionable category. See abh; au; aw; awni and owr.

outfal-drain (E) An agricultural drainage term. Vancouver (1808:286) states: "The outfal-drain being completed, and proper sluices erected to give a command of water ...." A term mentioned again on p 451: "Mr. Bayley, shall see the necessity of providing outfal drains by the ditches of his mounds ...."

outlet (E) The EDD (4:377a) gives: "a small channel or passage for water cut through the side of a road".

outlet drain (E) Rennie (132a, under covered drains) gives: "The dimensions of the conduit depends upon the quantity of water it has to carry; thus, in an outlet drain, it requires to be larger than in a cross drain, which has only the water collected in itself to discharge." Burke (1837:472) mentions the term too.

over-shot (E) Baker (1854:2:85) gives this term, the definition of which, points more to a fall of water in the mill tail rather than as a generic alternative: "The space over which the waste water flows, from the wear, down a sudden short declivity, to the natural course of the river. Frequently called a LOW-SHOT; and occasionally a STOOP."

ow (Ir., SGael? & Wel?) One of three contracted sound forms of abh – see Joyce (1869:3:54). The form ow is found as a common stem for river names in various counties; there is a river Ow in Wicklow; a river Owveg in Kerry and a river Owvane in Cork, many more are listed in GN. VI also gives ow as a generic term and further references can be found in Bartholomew (34a) and Blackie (2). Although Jamieson (supp. 1a) gives ow as one of the forms of abh, 'water', no particular instance of its use as a generic is given in SGael. Waddell (1927:2a) representing the Wel., also lists ow as "river, sea." See au and aw.

owein (AIr.) See owen.

owen (AIr.) One of the anglicized forms of abhainn, 'river'. See Joyce (1869:1:454-5). Used as a common prefix for river names throughout Ireland. On maps in particular, dual terms are very often shown for river and stream names such as Owenmore, the anglicised form, and Abhainn Mhor, the Irish form. The term prefixes a number of RN's on OSIDS 25 and 31.

owin(n) (AIr., Mx. & SGael.?) In Irish, owin is used as a variant of owen, 'river'. See Galway L for numerous instances of owin, owinn and a rare form, owein, used locally. For the Mx., Kelly (1866:1a) states that: "in ancient manuscripts a, o, and u, are written indifferently, one for the other, as clagh or clogh a stone; awin or owin a river", hence this spelling. Cregeen (1835) does not list awin or owin! For the SGael., Liddal (12) gives owin in a PN example: "Burowin. Barr + abhainn = summit of the river." The form Burvane, occurs also." See owen.

owle (E) See hohle.

owr (OCelt.) The Basque word for water *ur* has been suggested, at times, as a possible etymology for the Sc. RN Urr. Chalmers (48) states: "Ura, in Basque, is applied to water, a river." and further (50), that: "Or, Owr, in ancient Celtic, are applied to streams of water, and so is Ura in Basque." Maxwell (37) states that: "Mr Skene has drawn attention to the frequent occurrence of the syllable ll in the topography of the Basque province … there is perhaps more significance in the resemblance he traces between ur, the Basque word for water, and our river names Urr, Oure, Ourin, and Ore." With so many Or-; Our- Owr-; and Ur- RN stems found in northern Europe one would expect to find a common etymology but, up to now, no real connection has ever been made or a plausible etymology given. It's daring and controversial to include Bas. words, but with so many Ur- stems, the temptation is hard to resist; the foregoing suggests pursuing Bas. as it is now gradually becoming accepted as a possible early language arrival in the British Isles. See our; oure; *Notes on the Languages*, p xxiv, for the latest IE connection and especially, Vennemann, *Water all over the place: The Old European toponyms* *and their Vasconic origin;* and, on the Sc. RN Urr, Ross (219).

P

partition drain (E) An agricultural drainage term. Vancouver (1808:286) gives: "partition drains made three feet wide, and two feet and a half deep."

passag (Sc.) The DOST gives: "A water-course, gutter or conduit."

peat drain (E) An agricultural drainage term. Formed by using peat as the conduit, Stephens (5:263) gives a number of sketches.

pen (Sc.) Warrack (407a) gives: "an arch, archway; a small conduit."

penhead (Sc.) Jamieson (3:469a) gives: "The upper part of a milllead, where the water is carried off from the dam to the mill."

pennock (E) Parish (86) gives: "A little bridge over a water-course; a brick or wooden tunnel under a road to carry off the water." The variant form is given by Parish & Shaw (117) "PINNOCK. A wooden drain through a gateway."

penstock (E) A channel for water to drive a millwheel, given by Wikipedia.

pen-trough (E) Addy (43) gives: "The wooden or iron conduit by means of which water from a dam or reservoir is conveyed to the top of a water-wheel."

perrie-weerie (Sc.) Jamieson (3:462b) gives: "A slow-running stream."

peth (Sc.) The DSL (DOST) give this as: "watercourse."

pfifa (OHG) See pipe.

pidele (OE) KCD (3:xxxv) defines this term as: "a thin stream." It enters into the charters of BCS 120 and KCD 59 (S 78) and KCD 570 (S 786) and, Drayton's map of Dor., names the Trent: "Pidle"! For PN's derived from and retaining 'piddle', see PNWo (14, 155 & 222) and for names that have developed into 'puddle', see PNDo (1:288; 294; 309 & 311).

pill (E & Wel.) From OE *pyll*, 'tidal creek, pool in a river', but also "channels through which the drainings of the marshes enter the river." B-T (779). Although mostly found in Som. and Gloucs., pill can also be found in Oxon., at NGR SP2204 – The Pills – a junction of two streams near Filkins to which Archaeologia (145) refers. The plurality of 'Pills' here suggests that both streams were once called pill. A rare spelling of pill exists in KCD 654 (S 862), in the form: *"puylle – rihshammes puylle"*, now Rushen Gout, on this, see OSOL 14 at ST5990 and PNGl (3.118). Other pills in Gloucs. are shown on OSOL 14 at ST6699 and the PNGl has numerous references to pill, especially (4:164). Pills are common in parts of South Wal. too, such as the ones in the marshes of the Afon Llwchwr estuary, Gower Peninsula, shown on OSX 164 SN4700, and there are two on the River Taf estuary, Camarthen Bay, both shown on OSX 177 at SN2908 and 2909. See pyll; EPNE (2:75); PNBrk (1.112-3) and PNO (320 & 462). Other pills can be found in PNCh (1.281-2); PNSx (2.386) and PNW (444). Pilton in Rut., PNRu (289-90) and Pilland in Dev., PND (55) add yet more instances of this very widespread element.

pinmarch (Wel.) See pynfarch.

pinnock (E) See pennock.

pipa (E, OE & OFris.) See pipe.

pipe (E, OE & OFris.) From the OE *pipe*, 'The channel of a small stream', the word is listed in B-T (680) and is cognate with OSax., *pipa;* OFris., *pipe* and OHG *pfifa* as well as other Germanic languages namely Dan., Ice. and Swe. Lambarde (199) states: "Divers other smal pipes of water there be, that doe minister secondarie helpes to this Navigable River", and (260): "Wels (or springs) the which creepe at the first out of the earth, and bee conveied in slender-quilles, then afterwarde (meeting together in course) doe growe by little and little into bigger pipes, and at the last doe emptie themselves into some one bottome, and so make up a great streame, or chanell." Bannister (1916:153) under Pipe states: "It seems as if it must be O.E. pipe, 'a pipe'. A place called the Pipe, near Lichfield, is so called because the city water has for long been conveyed by pipe from there. But an explanation such as this could not apply to a Dom. name. Judge Cooke says the name Pipe is properly applied only to 'an elongated strip of land consisting of about 120 acres, through which *quasi per pipam* a stream known as the Pipe brook flows eastward to the Lugg." BCS 204 (S60) and KCD (118), in the bounds relating to a grant at Stoke Prior, both mention: *"Of þam bǣte in pipan"*. In addition to the Pipe Brook in Herefs., there is a Pipe Strine, shown on OSL 127 at SJ6918, which runs into Strine Brook and eventually the R Strine in Staffs. See EPNE (2:65); ERN (327); Hooke (68); the MED and PNC (340).

pipe drain (E) An agricultural drainage term given by Loudon (710:4296) and Johnson (128) who states: that: "The sod, or pipe drains, are undoubtedly the least expensive of any."

pipe-gutter (E) Elworthy (1886:573) defines this agricultural drainage term as: "A drain made with ordinary tile pipes, in distinction from a stone-gutter, which is one made of loose stones, until late years by far the commoner kind."

pirle (E) Leland (1:301) gives: "A broket or pirle of water renning out of an hille", and (2:37), "... issuith a little pirle." The EDD (4:520b) gives the form: "purle" from a quotation therein and the ERN (333) mentions pirle variants under prill.

pistol (Ir.) Joyce (1869:3:206 & 364) gives examples of this term from PN's – Cloghapistole and Glaspistol – defined as: "a half-hidden streamlet running in a deep tube-like channel."

pitch-gutter (E) Elworthy (1886:576) defines this agricultural drainage term, with a bit of local dialect added: "A channel or shallow open drain formed with small stones or pebbles. Thick road 'on't never be vitty gin there's a proper pitch-gutter a-put in both zides o' un."

*pleud. (IG) See fleet.

plug drain (E) Stephens (1889:5:262-3) gives a sketch of this agricultural drainage term and mentions that: "It resembled somewhat the wedge-and-shoulder drain when finished."

*pluti- (IG) See flood.

pol (Co.) As a generic term for stream, this word is not commonly attested in Co., the normal definition being 'pool', but, for a possibility, see CPNE (187-9).

poll (ME and OSc.) The DSL (DOST) connect it with the Wel., pwll and Co., pol, and define it as: "A slow-moving, ditch-like stream, flowing through carse-land." See pol; pow(e); EPNE (2:68-9); ERN (329); the MED under pol(e); PNCu (1:23-4 & 3:487) for a list of poll names and the CPNS (142, 204, 370 & 463).

pollan (SGael?) Only in Milne (1912:166) who gives: "Fowlis. Burn. Pollan, dim. of poll, pool, burn." The -an terminal is the correct dim. suffix, but, no other authority seems to support 'pollan' as a generic term. See poll.

pompe (ME) See pump.

pond (E) Not normally defined as a watercourse but is shown as such on OSX 134 at TQ1420 and 154 at ST5190.

ponsdowr (Co.) See dowrbons.

pont (Wel.) This term is listed in the GPC (2849a) defined, amongst other things, as: "aqueduct", first used in the 12th C. See dyfrbont, dyfrffordd and traphont.

poo (E) See pow(e).

pool (E & Sc.) A common watercourse term in the north of Eng., and also found in Scot. Ten instances of pool are shown on OSOL 6 (1998 edn.) and OSX 113 at SS7200 shows a Dev. one. Sc. pools are shown on OSX 438 at NH8484 and OSX 449 at NC8961. See The EDD(4:578b); the MED under pol(e); pow; pul; pwll and, on the whole subject, ERN (329-332) for a detailed and exhaustive account.

pou (Sc.) See pow(e).

pouran (SGael?) Only in Milne (1912:222) who gives: "Little Pourin. Small stream oozing from a hill. Pouran, small stream", and: (266) "Pourin, Burn formed by the drainings from a hillside. Pouran, small stream", the -an terminal suggests a dim. form, but, as with pollan; no other authority seems to support the fact that 'pouran' exists as a generic term.

pourin (SGael?) See pouran.

pow(e) (E & Sc.) Prevost (136) gives: "Pow" and "Powe" – "A natural sluggish, slow-moving stream, generally with a muddy bottom, the extent of water not being implied." Dickinson (306) lists: "poo, A wide and watery ditch." A pow can be seen on OSOL 5 at NY4649. In Scot., the term is much more widespread than in Eng. – Warrack (424b & 426a) gives: "Pou" and "Pow", defined as: "a slow running stream" in each case. Jamieson (3:536b) and The DSL (SND & DOST) also list pow, the latter under poll. A Sc. pow is shown on OSX 321 at NX9673. A lengthy discussion on pow and poll is given by Ekwall in ERN (329-332) who also lists the Cumberland watercourses given in the PNCu (1:23-4) and the CPNS (142) mentions the terms too. The Norf. Powdykes, shown on old maps, do not belong to pow. On this see the OED (7:1212c) and later editions.

prick-gutter? (E) Darlington (300) gives "Prick-gutter" and (409) "Trig-gutter" both as: "a small gutter." A definition which does not confirm their use as watercourse generics, therefore, they must remain questionable. However, Jackson (453) gives, for *trig* (2): "a small gutter, — same as Rigol", and defines rigol (352) as: "a small gutter or channel in land, made to lead water off." These definitions would make *trig-* , at least, a generic agricultural drainage term. See trig.

Priel (G) See prill.

prill (E & Wel.) Robertson (118) gives: "A little rill of water" – Jackson (335): "a streamlet of clear water, a rill" and Lewis (1839:82): "a small stream of running water." According to Bannister (1916:157) prill is: "A phonetic variation of O.E. pirle, purl, found only in Worcs., Shrops, Herefs., Rads. and Gloucs., a small stream of running water." There is a Prill PN, roughly halfway between Shrewsbury and Oswestry shown at NGR SJ3719. For the Wel., Evans (2:1858:886b) and Pughe (2:1873:418a) both define prill as a watercourse. Further confirmation is given in the GPC. See pirle.

puddle (E & OE) From the OE *pudd*, 'ditch'; a rare term, evidence is fairly thin, but a watercourse bearing this appellative is shown on OSX 123, lying north of Polegate at TQ5905. Morris (1857:44) gives: "piddle, puddle (Anglo-Saxon), a thin stream", which may have been taken from KCD (3.xxxv) where it states: "Pidele, *piddle*, a thin stream," and Pulman (1875:784) states that: "Purlbridge" is "Called by the "natives" Puddlebridge – *puddle* from the Anglo Saxon, for a *little stream*." The normal definition of puddle is 'a small dirty pool' usually formed by rain, but the evidence here suggests that 'puddle' has been taken in as a generic term for a watercourse, at least on the South Downs, confirmed by the modern map. See pidele and PND (622) for Purlbridge.

pul (Co.) Borlase (451c) gives: "a stream." Jago (1887:155a) lists pul, under stream, from Pryce's entry, but Pryce only gives "pit." This leaves Borlase as the singular authority. See pol and pwll.

pump? (E) The OED1 (7:1591b) states: "a pipe or conduit for conveying water." followed by a quotation from *The Medieval Records of a London City Church: St Mary at Hill, 1420-1559* – "paid to Mr Osborn ffor a pompe yat lythe to brynge the water owt of ye diche into ye ponde". The full text of which is viewable in the MED.

purle (E) See pirle.

pŵant (Wel.) This term is listed in the GPC (2936b) defined as: "mill-leat."

pwll (Wel.) The GPC (2940b) does not define pwll as a watercourse, the nearest being "ditch", in fact, neither does anyone else; definitions are always in the range, 'hole, ditch, pit, and pond' and the like, but, a good number are shown on the modern map as watercourse generics – two on OSX 240 at SJ2811 & 2911; OSX 213 at SN6289; OSX 240 at SJ3017 and OSOL 23 at SN6490. See pul.

pyll (OE) See pill.

pynfarch (Wel.) The glossary in the LL (li) lists: "Pinmarch=a pond, mill-race." Evans (2:1063b, under watercourse) gives the form "pynfarch" and in (2:341b) gives the compound Pynfarch melin, "millrace".

pynfarch melin (Wel.) See pynfarch.

R

race (E & Sc.) From the ON *rás*. Peacock (1869:66) gives: "Race, a small stream, a mill-lead." Mill Races are shown on OSX 131 at SU3421 and OSX 306 at NZ5723 and a Gypsey Race on OSX 301 at TA0272. A Sc. race is shown on OSX 321 at NX9395. See race- course; trough and PNWe (2:280).

race course (E) Brees (192) gives: "… the cut or canal along which the water is conveyed to and from a water-wheel." See race and race trough.

race trough (E & Sc.) This term is rare and comes from a letter written in 1822 by Thomas Carlyle, which can be viewed in *The letters of Thomas Carlyle to his brother Alexander: With related family letters:* Ed. by Edwin W. Marrs, Jr., at (Gbooks) or, in *The Collected Letters, Volume 2*, at the CLO website. The DSL (SND) mention race-trough, under offset. See race and race course.

racu (OE) See rake.

ragavon (Co.) George (2009:548b & 919b) defines this term as "tributary", it is cog. with the Wel., rhagafon. See avon vaga.

raibér (Ir.) See ruibér and eDIL.

rake (E & Sc?) From the ON *rák*, 'streak, stripe, crack, crevice', but, OE *racu* has a claim too. B-T (780) give: "racu, e; *f. A 'rake', rake a mountain track … a hollow path, bed of a stream*" and (927) the compound *stream-racu*, defined as: "*The bed* or *channel of a stream, a water-course.*" WW (178.5) also give *streamracu*, which is equated with the L *alueus* in the *Supplement to Alfric's Vocabulary*. A rake is shown on OSOL 41 at SD7061. The Sc. term relates more to 'a reach in a river' rather than as a generic. See alueus; EPNE (2:80) and PNYW (7:234).

ramblin-syver (Sc.) See rummle cundy.

rammel-cundy (E) See rummle cundy.

rang (Ir.) O'Reilly (414a) defines this term as: "a stream" and "the bank of a river."

rant (E) Charnock (1880:38) gives this rarity: "One day you may find snipes by the side of favourite rants and fleets in fair numbers (in the Essex marshes) ...." The EDD (5:36b) mention the term, which is referenced to the above.

rás (ON) See race.

rasán (SGael.) The only reference to this term comes from Dwelly (750a) who gives: "rivulet."

ray (E) Embleton (152) gives: "Ray, a stream." See rea.

re (ME) See rea and ree.

rea (E) This is the parent word of ME ree which came about accidentally. The phrase, *in thære éa* or *æt thære éa*, at some time in the past, was misdivided; the r was added to ea. This fact is pointed out by Duignan (1902:126) and PNSa (1:219). BCS 1007 (S 1185) is a good example of how a river can change name through misdivision: *"in thære éa nen"*, a quote from the charter, refers to the R Nen or Neen, a name which fell out of use leaving the misdivided rea as the new name. Although the RN was lost, Neen has survived in the PN's Neen Savage, Neen Sollars and Neenton, all in Shrops. The upper reaches of the River Rea through Neenton is actually marked Rea Brook on maps (SO6487) and there is another Rea Brook which falls into the R Severn at Shrewsbury. Birmingham has a R Rea too. In ME, the phrase, *æt thære éa*, often became *atter e*, *atte re* or *attere*. A typical example can be found in the 1327 Subsidy Roll for Staffs. (SSR 230), in this, "Ric'o Attoree" – 'Richard at the river' is listed as paying "12d" towards the Subsidy. Considering the foregoing, it's rather odd that only ree, and not rea, has survived as a generic term – rea and its variant ray is now only found as a common RN in the counties of Berks., Cambs., Oxon., Shrops., Wilts., Warks. and Worcs. See ea; ee; ree; ERN (336-7); the MED, under re and the relative PN volumes.

reach (E) This term is shown on OSL198 at TQ4508 and discussed in PNSx. (353). It is also 'a branch of a R', shown at NGR TG4705. Tipping (181) states: "Glynde Reach, flowing to the Ouse, a brackish stream on which at high water a boat of some fraught may swim. Local tradition makes much of this ritch or reach." A postulated OE *ric has been proposed by a number of authorities for Glynde Reach and the various Yorks. Skitterick PN's. In Wel, rhych and rych have the meaning furrow and water furrow (GPC 3125a). In the 13th C *Black Book of Carmarthen* (Evans 1906:33) we have the form "rich." The *Midderice* in BCS 814 and the *Beferic* and *Doferic* of BCS 1242 repeated in KCD 561 may have some bearing on the reason for the postulated form. See EPNE (83) and ERN (370).

rean(e) (E) From the OE *ryne*. Wright (789b) gives: "A gutter or watercourse"; (792B): "REEAN. A gutter" and (795b): "rene." Leigh (167) also gives reean and Cotgrave (under ruisselet) gives: "reane." See reen; rhean; rhin; rhine and rune.

red (Bret? & Co?) Lhuyd's entry (141b), under rivus, gives the Armorican (Breton) compound "dur red" and Henry (231) does not define red anything other than "course". Le Gonidec (358a) has the compound: "gwaz-red, torrent" and Bret V (197b), under rid, red, gives: "course" and "deur-rid, eau courante, ruisseau" The red form corresponds with the Wel., rhed and rhedfa, but, as a watercourse generic, it is not so well attested in Bret. and Co. as it is in Wel. – very few references exist – for the Co., one is given by CPNE (196) but, other than that, evidence, supporting red as generic, is lacking. However, in view of the above, red is not so questionable in Bret. as it is in Co. See goeth and the rhed- forms.

ree (E & ME) This is a development of rea by misdivision of the phrase *æt þære éa*, in ME this became *atter e*, *atte re* or *attere*, as in the name Attoree – see rea. The word appears in various historical works of note such as Holinshed (1:79) who has a classic line in one of his chronicles: "Rhe, or ree, the Saxon word for a water-course or riuer; which maie be seene on Oueree, or Southeree, for ouer the Ree, or south of the Rhee, as to the skilfull doth readilie appeere." There are many other references which relate to re, ree or rhee – a Battle Charter (112) dated 1451 mentions "Le Ree" and Furley (178-200) defines rhee as a Saxon word for a watercourse also; which he comprehensively coverers in his paper, *"An Outline of The History of Romney Marsh"*, as does Canon Scott Robertson (261-280) in *"The Cinque Port Liberty of Romney."* Willis (1:212) contributes: "Mylnestrete to the water called Le Ree" from his *Architectural History of the University of Cambridge* and other name and PN examples can be found in the MED under re. The OED1 (8:320c) suggests a connection with Flemish reie and rui – Kiliani (432a & 447a) lists reye and ruye respectively and there is a R Reie in Belgium which would support the suggestion. Taylor (206-7) has an interesting summary on the RN Rhe but his work, apparently, is not taken too seriously. See ee; rhee and the MED, under re.

reean (E) See rean(e).

reen (E) A variant of rhine. Jennings (1869:52) gives: "Reen" and "Rhine", a watercourse; an open drain" and Robertson (126) defines it: "A small brook or broad ditch, 2. The deep furrow between the "ridges", to carry off the water." Heslop (571) for the northern counties also gives reen. Mapped instances of reen on E maps would appear to be non existent, but there are plenty to be seen on OSX 152 in Wal. See reane; rhean; rhin; rhine; rune and waterrene.

ren (E) See rhine.

rendylle (ME) See beck and rindle.

rene (E) See rean(e).

*renno (Gm.) See runnel.

*rennon (Gm.) See runnel.

rennyng (ME) See running and wissing.

res (Co.) Nance (1978:287b) gives: "course, race, running of water", confirmed by George (2009:556b). It corresponds with the E race, which exists in this form as a Co. PN – see Weatherhill (59a).

reuer (ME) See river.

reuma (L) WW (1:183.8) give: "gytestream" as the OE corresponding term, from the *Supplement to Alfric's Vocabulary*.

revel(le) (ME) See rivel.

revelling (E) See riveling.

revver (E) See river.

revyre (ME) See river and rivus.

reyne (E) Randle Cotgrave, in his *Dictionarie of the French and English Tongues*, published in 1611, gives, what appears to be the only use of the word reyne, in the sense 'brook'. Under ardoüe he gives: "a little brooke or reyne", a spelling repeated in his entry for riu: "a little brooke, a reyne, or gullet of water." Further, under seillon, he gives: "the narrow trench, reyne, or furrow, left betweene butt and butt for the drayning thereof." So, we have here, a rare, now obsolete spelling, which has not survived beyond the 17th century as a generic term. See the MED, under rein.

rhaeadr (Wel.) The GPC (2997a) gives: "waterfall, cascade, torrent" – two can be seen on OSX 215 at SN8193.

rhagafon (Wel.) Cog. with the Co. ragavon – Spurrel's (323b) define this as "tributary", GPC (3000c) list it too. See afongainc.

rhe(e) (ME) See rea and ree.

rhean (Wel.) Evans (2:886a & b) gives: "a small stream and streamlet" respectively. Compare the rhen and rhin terms and the E forms.

rhedfa (Wel.) Listed in the GPC (3045b & c) and defined as: "watercourse, channel; slope down which water flows", a term, dating back to 1794. See red.

rhedle (Wel.) Spurrell's (1934:326b) gives: "course, channel" and Pughe & Pryse (1:655a, under dyfrle) give the compound rhedle afon: "the channel or bed of a river".

rhedle afon (Wel.) See gwely afon and rhedle.

rhedlif (Wel.) Pughe & Pryse (1:243a, under awon) give: "a stream; a river."

rhedweli (Wel.) This term and its variants: rhwydweli and rhydweli are all listed in the GPC (3046a) defined as: "drain; river-bed, watercourse, channel" in use since the 15th C. See red.

rhein (E) See rhine and wring.

rhen (Wel.) Pughe & Pryse (2:446b) give: "a brook, a rivulet – ffrwd". See rhine and rhyn.

rhewin (Wel.) Pughe & Pryse (2:447b) give: "a little gutter" and also list the variant "rhewyn". Walters (1:169b) gives: "a little brook".

rhewyn (Wel.) See rhewin.

rhidys (Wel.) Evans (2:711a, under rivulet) gives: rhidys and its dim. rhidysen.

rhidysen (Wel.) See rhidys.

rhigol (W) The W form of the E rigol. Pughe (468a) gives: "a grove; a trench, a furrow; a drill; a small ditch, a drain."

rhin (E) See wring.

rhin (Wel.) Pughe (1832:2:468a) gives: "a channel which carries off lesser waters; in draining a name for the secondary dykes, which receive the small ditches, and convey the water to *gwyth fawr*, or great channel; a furrow between lands." Pughe & Pryse (under ystrym, 2:636a) state "also called a rhin". See rhen, rhine and rhyn.

rhine (E) Probably from the OE *ryne*, A common term for a drainage channel on the Somerset Levels, but well recorded in Gloucs. too. Jennings (1869:52) and Dartnell (132) both give rhine and Smiles (1878:112) has: "rhein"! Pulman (1871:130) lists "rine"; Hill (329-30) gives: "ren, watercourse" and Bailey has: "ROYNES [in old records] Currents, Streams or passages of running Water". Other forms can be seen in The OED1 (8:629a). Two forms exist on maps – rhyne is common SW of Bristol, on OSX 141 at ST4353 and rhine is common N of Bristol on OSX 154 at ST5682. See reane; reen; rhean; rhin; rune and the PNGl (3:114, 131 & 138, and, especially, 4:166) for watercourse names.

rhwydweli (Wel.) See rhedweli.

rhych (Wel.) See reach.

rhych dw(f)r (Wel.) Evans (2:1063b, under water-furrow) gives: "rhych dwfr and the variant rhych dwr."

rhydweli (Wel.) See rhedweli.

rhyn (Co.) This term and its variant forms, rin(e); ruan and ryne are only found in the lists of Dexter; Jago (1887); Pryce and Williams (1865). Dexter (1926:38.97) gives: "rhyn" as "river

channel" and Pryce lists: "rin(e); ruan and ryne, the channel of
a river." The Jago and Williams entries follow Pryce. Harvey
(102) quotes "The Reen" from the Tithe Award with the true
Co. spelling "ryne". This form is found in A-S; B-T (806a)
define it as: "*a course, watercourse*", but, like the guy variants,
none have found their way into a current Cornish dictionary.
See reen and rin(e); George (2009:560a rhine) and Norris (477-
9), who discusses the Pryce forms rin(e); ruan and ryne, and,
on the ryne form in particular, PNGl (4:167) and PNYW
(7:239).

rhyne (E) See rhine.

*ric (OE.) See reach.

rich (Wel.) See reach.

rid (Bret.) See red.

ride (E) From the OE *rið* WW (1:177.38) and *rip* B-T (689, 800)
and also recorded in OFris. Cope (74) and Grose (130) both
give: "a little stream." Ryde, in the IOW, is probably the most
well known PN linked to this element and there are many
others with the rið or rip element. See rife; rithe and EPNE
(2:85-6 for lists of PN's).

rie (LG) See rife.

riefa (LG) See rife.

rif (Dut.) See rife.

rife (E) From the OE *rip*. Kiliani (435b) gives the Dut. cognate
"*rif*", which may be associated with the LG *riefa*, "fluss, river"
and the contracted form *rie*, "creek, brook." Parish (96) gives:
"A ditch on the moorland", and it is a very common name for
drainage channels shown on OSX 120-1. See ride; rithe and
PNSx (1:4, 73 & 83).

rigatt (E) See riggot.

riggat(e) (E) See riggot.

rigget (E) See riggot.

riggot (E) Jackson gives: "rigot, same as rigol." Wright (802a)
has: "Rigatt, a small channel from a stream made by rain.
North" and Darlington (315) states: "riggut, a channel, gutter."
There are many forms of the word – Addy, for the Sheffield

region gives: "RIGGAT(E), a small watercourse or stream", and mentions that: "Riggot occurs as a surname in Dronfield" – Evans (225), in his *Leicestershire Words*, gives: "Rigget, a small water-furrow or surface drain." The EDD (5:105b) furnishes us with "wriggate" and, in addition, Grose, Leigh, Nodal and Wilbraham all have entries covering this very widespread element. See rigol.

riggut (E) See riggot.

rigol (E & OF) From the OF *rigol*, 'canal'. Jackson (352) gives: "a small gutter or channel in the land, made to lead water off." Still used in Bel. and Fra. for irrigation and drainage. See USBG (Bel. & Fra.).

rigot (E) See riggot.

ril (Fris. & LG) See rill(e).

rill(e) (E, G & LG) 'A small steam, brook or rivulet'. A term which, according to the OED1 (683c), corresponds with: "Fris. ril, LG. ril, rille" and "G. rille". Skeat (1882:519) says that it is also found in the Norman dialect and Leland (1:36) gives: "There is a rylle that cummith by the towne ..." and (90): "rille". Holinshed (1:96) states: "a little beneath meeteth with a rill ..." and Drayton's fourth song gives: "The Bourns, the Brooks, the Becks, the Rills, the Rivelets." A mapped example can be seen on OSX 150 at TR2357. See PNWe (2:281).

rillet(E) A dim. of rill, Wright (1880:802b) gives: "a small stream, a rivulet." Leland (3:73), during his tour of Wal., states: "There cummith a litle ryllet by this square ...." He also gives rylletts on the same page. Drayton (1:10, song 1) adds another: "Those rillets that attend proud Tamer and her state ...."

rillock (E) A dim. of rill, Rotzoll (159) gives: "a small stream." Gilchrist in his book, *The Peak District*, (51) mentions the word too: "There, from a gloomy ravine called the "Salt Box", a rillock creeps and soon loses itself in the grass." The sentence here implies that a rillock is nothing more than a trickling streamlet.

rin (Co., E & Sc.) For the Co. see rhyn. For the E, Wright (1880:803a) lists rin as an A-S word and Heslop (578) gives:

"rin, a small stream" and (590) "ryn, a very small stream." B-T do not list rin(e) only ryne, as a watercourse (806a) and Clark Hall equates rine (244a) with ryne (245b) but, not as a watercourse. Just to add a little confusion, Pulman (1871:130) gives: "Rines, The drains in the Somersetshire moors are so called", used here, no doubt, as a variant of the E reen, rhine and rhyne. For the Sc., Warrack (457b) defines rin as: "a small stream" and there are many entries for rin and its compounds in The DSL (DOST & SND) worth perusing. See rhyn and all the rin- compounds.

rind (E) A contraction of rindle perhaps. Not normally defined as a watercourse but is shown as such on OSL 110 at SK1188 as the name of the upper reaches of Blackden Brook.

rindel (E) See rindle.

rindle (E & Sc.) From the OE *rinelle* and *rynel(e)*, B-T (689, 691, 800 & 806). Balg (325b & 326a) in his *Gothic Glossary* under "*rinno*", states that it is: "allied" to this term and "runnel". Bosworth (546a) gives: "*rinnon*" from John (18:1) and Köbler (Got.), allies *rinno* to "runnel" and "brook", but not rindle. It is recorded in the *Vespasian-Psalter*, c825, and listed in the glossary by Grimm (168), in the form, *rinnelle* and the ME form "*rendylle*" is given in the PP (29). Darlington (315) gives: "a rivulet" and Jackson (352) has: "rindel, a small stream." For the Sc., Warrack (458a) gives: "rindle" and (459a) rinnal. Jamieson (Supp., 204b): "rinel and rinnel, a runlet, gutter." The DSL (DOST) has "rinel" and The (SND) "rinnal." See runnel; The EDD(117a); the MED, under rinel; PNDb (4:747); PNWe (2:282); PNYW (7:239) and PNSa (6:78) for Rindleford.

rindlet (E & Sc.) The dim. of rindle "a rivulet." The EDD (117a) gives: "RINDLET, a rivulet" and (121b): "RINLET, A small stream." Warrack (459a) for the Sc., gives: "rinlet, a small stream." See Rotzoll (159 & 160).

rine (Co., & E) See rin.

rine (E) See rhine.

rin(n)el (Sc.) See rindle.

rin(n)elle (OE) See rindle.

rinlet (E & Sc.) See rindlet.

rinnal (Sc.) See rindle.

rinnand (Sc.) The SND (DOST) gives: "watercourse."

rinnar (Sc.) See rinner.

rinnek (SNn.) Jacobsen (1928:700b) gives: "a small brook."

rinner (Sc.) Warrack (459a) gives: "a stream, brooklet." Jamieson, in his single volume *Dictionary of The Scottish Language* (437b) gives: "a little brook." The SND (DOST) list this form under rinnar.

rinnick (OrkNn.) Marwick (134a) gives: "a small drain or cutting by which to drain away water." and the variant (138a): "runnick".

rinnie (Sc.) A dim. of rin. See rin-water.

rinno (Got.) See rindle and runnel.

rinnon (Got.) See rindle.

rin water (Sc.) The DSL (SND) give: "a natural flow of water, esp. one used to drive a mill without the necessity of a dam."

riolet (A-N) Moisy (871) gives: "petit ruisseau", for this term, 'rivulet'."

riparia (L) Martin (310) gives: "riparia: - water flowing between banks." See ripary.

ripariam (L) See ripary.

riparie (ME) See ripary.

riparii (L) See ripary.

ripary (ME) From the L *riparia – ripa*, 'river bank', a very rare term for a watercourse given in the OED1 (8:702b). In Black's Law Book (1041b), it is defined in different ways: "RIPARIA. A medieval Latin word, which Lord Coke takes to mean water running between two banks; in other places it is rendered "bank"; Martin (310) gives the same: "riparia: - water flowing between banks." L words with the ripa- or ripe- stem are all to do with the river bank or edge and the like, for instance, a riparian owner is one who holds particular rights to the water that passes through their land which could be used for supplying a water mill for example. Ripary or Rivers, as a surname, is given in a number of volumes of state and elsewhere; recorded in the time of Henry III (1216-72), and earlier, as Ripariis, the common form in Fees (93) and Riparia (236:notes). Also in Fees (480), we find River and Ripariis as variants of the same name. AD has many entries for the name too. A line from Allen (723) states: "Richard de Ripary, i.e.Rivers" and again, in Newbury (275), we get mention of: "... the family of De Riparies or Rivers." It is obvious that, on occasions, one form is substituted for the other. The *Riparies* of De Riparies is the same spelling as that given in Godstow (559, line 17). Ripera, the Latinized form of the PN River in Kent (Fees 269), is obviously related to riparie; a variant of it. There is a Rio Ripary in Spain, given in the notes on Ravenna (288) as: "Dora Riparia." Some geographical works, such as Forbiger (3:366), give the name Ripera, for this R, the same form as the Kent name! The foregoing suggests that, although Black's: *'water running between two banks'*, is questionable, support for it can be found in the AD (4:279:A. 8231) which gives: "...the stream (ripariam) ...", in a deed dated 1261/2 and (6:118:C. 4688) "... of the river (riparii) ..."; two entries showing that *ripariam* and *riparii* were used as the L form of stream and river respectively at this time. All the forms given herein, adapted from L riparia, in the latter part of the Medieval period (generally considered to fall between the 10th and 15th C) are variants of each other, used to define a generic term in the ME period. See the MED and river.

ritch (E) See reach.

rit-fure (Sc.) The EDD (5:128b) gives: "the first furrow opened in ploughing; a furrow to run off surface-water in a ploughed field."

rith (OLFrk.; OFris. and OSax.) See rithe.

rið (OE) See ride, rife, rithe and rye.

*rīþa- (Gm.). See rithe.

*rīþaz (Gm.). See rithe.

rithbhir (Ir.) O'Reilly (424b) defines this term as: "a river." apparently from "uisge reatha" the Irish equivalent of the E 'running water'. Most Irish words beginning with reath- are all to do with 'running or strolling'.

rithe (E) Another term inextricably linked to the OE *rið* and *rið(e)*, B-T (800). Cognates are: OLFrk.; OFris. and OSax., *rith*; OHG, *rid*. The Gm. forms are **ripa-*, **ripaz* and the IG is **er-* and **or-*, all from Köbler. Parish (97) gives: "RYTHE, A small stream." Three rithes can be seen centred on SU7400 and there is a R Rythe in Surrey shown on OSX 161 at TQ1565. The Wrythe, an area marked on OSX 161 at TQ2764, north of Carshalton, Surrey, belongs here too. See ride; rife; the MED; PNSr (42); EPNE (2:85-6) for a discussion and lists of PN's, and ERN (342). See rye.

riþ(e) (OE) See ride, rife and rithe.

riðig (OE) B-T (801) give: "*a stream*". A term which does not appear to have come down to us in ModE. See EPNE (2:86); PNO (2:419) for PN forms and KCD 620 for charter forms.

riual (L) See rival.

riuel(l)et (E) See rivelet.

riuer(e) (ME) See river.

riueret (E) See riveret.

riuierette (E) See riveret.

riuilo (L) The CLB (227) gives: "riuilo", from a Leet Order dated 1446.

riuulo (L) See rivulo.

riuulum (L) See rivulum.

riuulus (L) See rivulus.

riuum (L) See rivum.

riuus (L) See rivus.

rival (E) A rare term which, in its Latinized form as riual, is found in a *A Decacordon of Ten Quodlibeticall Questions Concerning Religion and State*, by William Watson first published in 1602, and on p 68 of the republished edition by Scolar Press in 1974 (GBooks). See the OED1 (8:717b).

rivalet (E) This is a dim. of rival; found only in Lediard (277) "... being advanced to pass the rivalet ...." According to OED 1 (8:717b) an "obsolete form of rivulet."

rive (E) The OED1 (8:717C) gives: "a stream or rill", and the

ME variants: "ryve and riue." Battle (139) gives: "Mylryve".

rivel (E) OED1 (8:719b) gives: "rivel, A rivulet." Wright (1880:805a) gives: "RIVELLE, A rivulet", Halliwell (680b): "revelle" and the MED gives revel as: "A stream, brook." See PNYW (7:236).

rivelet (E) A dim. of rive. Drayton (3:175, Song 27) gives: "Which Spodden from her spring, a pretty Rivelet ...." and Holinshed (1:69) lists: "riuellets" and (1:95) "riuelet". Wells (2:426 & 428) gives: "... rivilets...."

riveling (E) A dim. of rive. Drayton (3:193, song 28) gives: "And Willowbeck with her, two pretty Rivellings." Saywell (78) gives: "revelling". See PNCu (371 & 488); PNDb (4:747); EPNS (2:86); ERN (343 & 401); Goodall (242); PNYN (7) and PNYW (7:136, 236).

rivelle (E) See rivel.

rivelling (E) See riveling.

river (E, Ir., Mx., NIr., Sc. & Wel.) From OF *riviere*; ModF *rivière* – from LL *riparia* – L *ripa*, 'river bank' – ML *rivera* and *riveria*. A term for the largest body of flowing water in the watercourse hierarchy, commonly found throughout the British Isles – every region has its 'river'. Cognates are: Pg., ribeira; Sp., ribera; Ital., riviera and M and ModDut., riuier, (Kiliani 438b). The LG form is: "Riwa", 'river'. Martin (310a) gives: "riparia, water flowing between banks"; "rivaria, a river bank" and "rivera, river." In Black's Law Book (1041b), riparia is defined in different ways: "A medieval Latin word, which Lord Coke takes to mean water running between two banks"; in other places it is rendered "bank" (see ripary). Other etymologies vary between 'shore' and 'river bank'. In Eng. there are many forms of the word, particularly in dialect – *revver* is just one; the IOW form, given by Smith (1881:28). The Irish Republic; the anglicized Northern Ireland Counties; the IOM; Scot. and Wal. too have all seen major changes, here, river has replaced many of the indigenous terms. As mentioned in the *Notes on Maps*, and repeated here, the 17th century designation for the majority of Irish river terms, shown on old maps, was 'flu' and 'water', much the same as the terms found on E maps, first and foremost, because the Ir. survey was carried out by E cartographers, who failed to take the Irish generics, prevailing at that time, into account. In fact, the frequency of river on the OSNIDS 7 map shows just how common it is in NI and OSIDS 31 shows its common occurrence in the Ir. Republic. In the IOM, the frequency of river just about exceeds every other watercourse generic term on the 1/25,000 maps; most of the indigenous terms have all but disappeared with the exception of awin, strooan and struan, and, a few other terms which have survived in PN's. In Scot. river is common in lowland but has also penetrated the Gaelic regions. Wyntoun has a number of entries for 'rywere' in his *Orygynale Cronykil of Scotland*. The Wel., too, now have many 'rivers', especially borderland, although *afon* is still the dominant term in the majority of counties. Salesbury, in his A *Dictionary in Englyshe and Welshe*, originally published in 1547, gives: "abe ne afon, A ryuer." WW (1:736.23) give the ME: "revyre"; the CDME (193b) gives the ME variants: "ryuer(e) and riuere" and the CLB (170) gives: "reuer" – Holinshed frequently uses the riuer form with the L, u for v, current in his time. River, a village in Kent (NGR TR2943), situated on the R Dour, is probably the only place in the country with a 'river' derivation. An entry dated 1219, in *The Book of Fees* (269), gives: "Ripera". See DEPN (371a) for the Kent PN, and, especially ripary. Over one hundred different river forms are given in the collective volumes of The DSL (DOST & SND); EDD; MED and OED and many other variants are listed in word and PN book collections too.

rivera (L) Martin (310a) gives: "river."

rivere (A-N) Another variant from the F *riviere*, given by Moisy (872). It enters into ME and many other variants are listed in The AND under this form.

riveret (E) A dim. of river. Drayton (1:45, song 2) gives: "A riveret born of her ...." Wright (1880:805a) adds: "A rivulet." Holinshed (1:102) gives: "riueret" and Cotgrave: "Riverotte. A brooke, little streame, small river." In addition, Leland (3:19) gives: "ryveret", a form fairly common throughout his *Itinerary* and Palsgrave (1852:240) presents us with what is probably the longest form of this word: "riuierette."

riverlet (E) A dim. of river, this word is used 4 times in *Romantic Richmondshire*, p 193, by Harry Speight and also features in the novel *Belinda*, p 60, by Rhoda Broughton, "Here by the riverlet sits the floury mill, and past it the quick stream runs...."

riverling (E) A dim. of river. Sylvester (1:78b line 755) gives: "riverling". There is a R Ling in Sc., shown on OSX 429 at NG9533, which might belong here, but according to MacBain (1922:62) the R Ling: "is in Gaelic "Abhainn Lumge, "Ship's River," connected with Loch Long."

riverotte (E) See riveret.

riviere (F) See rivere.

rivilet (E) See rivelet.

rivilum (L) The CLB (208) gives: "rivilum, from a Leet Order dated 1444.

rivo (L) AD (6:199) gives: "two brooks (*rivos*)".

rivola? (L) Martin (310a) gives: "stream, a rivulet." It may be related to the Italian *rivo*, *rivoletto* or *rivolo*, dims. of L, *rivus*. All mean 'brook, small stream, rivulet', otherwise it remains questionable.

rivulet (E) A very common term used especially in poetical works, from the L *rivulus*, dim. of rivus. Ogilvie (1:718c) gives: "A small stream or brook; a streamlet." See OSX 131 at SU4151 for a mapped example and the OED1 (8:723) for multiple forms of this term.

rivulo (L) BCS 673 (S 1552) gives: "rivulo" – KCD 355 (3:408): "riuulo", here, referring to the "Tamyse"; the R Thames.

rivulum (L) AD (6:528 C. 7838) give: "the brook (*rivulum*)", and KCD 140 (1:169), (S 116): "riuulum".

rivulus (L) A dim. of rivus, together, these terms have become the ones used in Mod L to define the smaller watercourse generics; flumen, fluvius and, to a lesser extent, amnis, are used for the larger streams and rivers. It might be expected that the

E term river, goes back to the L rivulus forms, but, that has a different root. Ogilvie (3:718 under rivulet) states: "L. *rivulus*, dim. of *rivus*, a river. 'a small steam or brook; a streamlet." WW (1:178.6) give: "*Riuulus*, lytel rid." and (736.24): "Hic rivulus, a bek", from two of the vocabularies. KCD 1154 (5:301) and BCS 810, (S 517) give the same.

rivum (L) AD (5:41:A 13574) mention: "... a brook (*rivum*) ...", and Earle (1888:330) gives: "riuum", (S 387).

rivus (L) WW (1:736.23), from a *Nominale of the 15th Century*, give: "*Hic rivus*, a revyre"; E 'river', and (177.38): "*Riuus*, rid"; E 'brook' from the *Supplement to Alfric's Vocabulary*.

riwa (LG) See rife and river.

ron(n)ek (SNn.) Jacobsen (1928:714a & b) gives: "ronek, ronnek, runnek. a small watercourse, brook."

royne (E) See rhine.

ruan (Co.) See rhyn.

rubble-drain (E) An agricultural drainage term. Apparently, 'a tributary drain'. Loudon (708) states that: "The common rubble drain is formed of rough land-stones of any sort, broken so as not to exceed two or three inches in diameter." Arkell (325) gives this term in his paper *On the Drainage of Land*, stating that: "Stone drains are various; the most common here are wall and dribble, or rubble, the former as main, the latter as tributary." See dribble, stone and wall drain.

ruibéir (Ir.) See ruibér.

ruibér (Ir.) This term together with its variant ruibéir appear in Earls (70, line 22 & 23). See raibér.

ruissel (E?) This term appears in the History of Jason (173), in the form ruysseaul, a 15th C work printed by William Caxton: "two ruysseauls or two springes of a fontayne", given by the OED1 (8:881c). According to Wiktionary, the word is really OF, from the "Vulgar L (c200-900 CE) *rivuscellus*, dim. of Classical L (c100 BCE -200 CE) *rivus*". Caxton's work would appear to be the only recording of the term. See the MED, under breken.

ruith (SGael.) Dwelly (778b) defines this term as: "Flowing, act of flowing as a stream", so here, we have a case of a verb becoming a noun as a ruith is shown on OSX 386 at NN8151. Milne (1912:278), under the PN Rootie Linn, also gives the same form and Dwelly (995b) gives the tautological compound: "uisge-ruithe, running water, stream".

rumbel drain (Shet.) Angus (111) gives: "a drain made by rumbelling stones into a trench and covering them with earth."

rumbling drain (E) An agricultural drainage term given by Loudon (581) and Burke (1841:20).

rummle cundy (E) Palsgrave (39) gives: "a ditch filled up with loose stones, for water to drain through." Heslop (588) has: "RUMMLE-CUNDY" and "RUMMLIN-CUNDY". The EDD (5:25a) gives another variant: "rammel-cundy" and the OED1 (8:890c) give the Sc. variant "rummle drains" in the same sense. Other Sc. variants with the same meaning are given by Warrack (441a): "Ramblin-syver" and (469a): "Rummlin-sive."

rummle drain (Sc.) See rummle cundy.

rummlin-cundy (E) See rummle cundy.

rummlin-sive (Sc.) See rummle cundy.

run (E) Peacock (1889:450) gives: "A small channel of water; a *runnel*." There is a run marked on OSOL 40 at TM4699. Other runs are shown at NGR TG3010; TG3209 and TG3307. See rin; PNBrk (3:786, 901).

runaway (E) Rye (181) gives: "At Wisby there is an open ditch across the green where the water runs across, called the *runaway*."

rundel (E) See rundle.

rundle (E) A variant of runnel. Holland (295) gives: "a small running stream" – Leigh (173) "a small brook, a runlet" and Holinshed (1:79) uses the form rundel. A rundle can be seen on OSX 274 at TF4475 as well as the tautological rundle beck at SK757324. See PNNt (266).

rune (E) From the OE *ryne*. Phillips (1720) gives: "Rune, (Sax.) a water course, so called in the marshes of *Somerset* shire" and Wright (1880:815a) states; "A water-course. West." Hill (131) has much the same: "it seems the abbot of Glastonbury had choked up certain watercourses called runes." See reen; rhine and PNBrk (2:531-2).

runlet(t) (E) Wordsworth (29) gives: "RUNLET, a water-drain", and, in addition, quotes an 18th C document: "Paid Herbert for two days Work at scowring Wire Lane Runlett, 20 June, 1755, 1s. 6d. Parish Accounts." Pegge (59) gives: "a small stream." See PNYW (2:11).

runnek (SNn.) See ron(n)ek.

runnel (E) According to OED1 (8:912a), a later form of rindle. Atkinson (1868:419); Dickinson (269); Ellwood (1895:52) and Robinson (1855:144) all mention this term variously defined as 'stream, drain, gutter, rill, or runlet'. Köbler allies *rinno* to "runnel" and gives the Gm. forms **renno-* and **rennon*, and, the IG **ere*. A runnel is shown on OSX 286 at SD4930. See rindle.

runner (E & Sc.) Prevost (269) gives: "A small stream." For the Sc., Warrack (470a) states: "a small channel for water." Two runners can be seen on OSX 285 at SD4119 and a Sc. one is shown on OSX 444 at NC7920.

runnet (E) The OED1 (8:913c) gives: "A stream or river; a runnel."

runnick (OrkNn.) See rinnick.

running (E) The OED1 (8:914b) gives: "A channel or watercourse; a stream or rivulet. Somewhat *rare*." The ME form is *rennyng* also given in the OED1 from the *Cursor Mundi* (line 11942). For PN evidence see PNGl (2:77) and PNWe (1:185).

running ditch (E) Rye (63) gives this term in his definition of drain: "a rivulet or running ditch."

runway (E) The OED1 (8:917c) gives: "The bed or channel in which a stream runs."

rut (E) Not normally defined as a watercourse but is marked as such on OSOL 21 SD9230. An appropriate term, as rut is defined as 'a groove or channel' – here, obviously a channel for water.

ruysseaul (E?) See ruissel.

rych (Wel.) See reach.

rye (E) This points to a contracted variant of OE *rið*, 'streamlet'. Holwell rye, near Cheddar, Somerset at ST4453, has rye as a generic term. A rye in Surrey is shown at TQ1658, and there is another stream, The Rythe at TQ1565. Peckham Rye was "Peckham Ry" in 1589 (AD 3:215) and the Wrythe, an area marked on maps north of Carshalton, belongs here too. On these courses and other PN's see PNSr (5, 21, 42 and 347). The MED, under 're', questions a watercourse definition in the first instance; most other forms in the MED relate to ree not rye, but the evidence here suggests that rye is a valid term. See rithe.

ryefere (A-N) Riley (1859:1:466) gives: "Le meilloure malard de ryefere pur iiid" and, in (1859:1:288), he gives: "le ryvere de Thamise". In A-N times 'Mallards from the river' were always worth much more than Mallards from a pond. Basically, although the forms given here are A-N, they are just one of a great number of variant forms from the O and MF which, ultimately, go back to L. See river.

rylle (E) See rill(e).

ryllet(t) (E) See rillet.

ryn (E) See rin.

ryne (Co.) See rhyn.

ryne (OE) See rhine and rune.

rynel(e) (OE) See rindle.

ryu (A-N) Kelham (211b) gives this and ryz and states: "a brook."

ryuer(e) (ME) See river.

ryver (Co.) Nance (1978:289a) gives: "river, stream." Borlase (453c) gives the Armorican (Breton) form "rywier, a river", quoted by Bannister (1871:138b).

ryver(e) (A-N) See arm and ryefere.

ryveret (E) See riveret.

ryvire (A-N) From "ryvires" in Kelham (211b).

rywere (Sc.) See river.

rywier (Bret.) See ryver.

ryz (A-N) See ryu.

S

salus (OPr.) See hail.

saluze (Su.) See hail.

saugh (E) See sough.

savig (Co.) A rare term, found only in Borlase (454a) defined as: "the branch of a river" –probably, the Co. equivalent of the E grain.

scaig (Sgael.) See ascaig.

schugh (Sc.) See sheugh.

scoor (Sc.) See score.

score (E & Sc.) From the ON *skor* 'a cut'; a rare term, but we know it to be one from the quotation given in Bridlington (143) which states: "the fosse called Syrithescore towards the west." Note 4, on p 144, mentions: "and 'sygridscore' is now written instead of Syrithscore." A foss is a watercourse term from the French *fosse* 'ditch, drain' and, ultimately, from L *fossa* 'ditch'. Another document in Bridlington (40) states: "a fosse for bringing water from Ruddestain", therefore, the case for 'score', as a watercourse, is proven. For the Sc. Warrack (482a) gives: "Scoor, a run of water; a channel."A score is shown OSX 329 at NS9811 in Scot., but, as far as I am aware, it is not shown on English maps, further, it does not seem to be mentioned as a watercourse generic in word books outside of Scotland. See PNYE (328) and YW (7:245).

scurf (E) See skerth.

secket (E) See sicket.

seech (E) See sike.

seek (E) See sike.

*seik (IG) See sike.

sele (E) Heslop (616) gives: "a marshy watercourse, a stream creeping through reeds and rushes."

seoh (OE) See seohtre(s).

seohter(es) (OE) See seohtre(s).

seohtra (OE) See seohtre(s).

seohtre(s) (OE) B-T (864) give: "seohtre, sihtre, *A pipe through which a small stream is directed, a drain*." Middendorf (116) gives the following examples under seohter, all taken from OE charters and all verifiable: "*seoh; seohter; seohteres; seohtra; seohtres; sihter; sihtran and sihtre*." He also mentions the E Fris. '*sichter*' used of a small stream for which Jellinghauss (116) gives: "wasserinne", 'waterway, channel' and, (35): "Sichtigoor". Pokorny (3:893) gives: "sichter" and "sechter". Two BCS charters use the sihtran form; 361 (S 1597) gives: "*sihtran*" and in 233 (S 126) we have: "in þætan *sihtran* of þam þætan sice ...", here, *sihtran* is synonymous with *sic*. More forms are given in other charters – BCS 50 (S230) has: "*seohtra*"; BCS 877 (S 552): "*sihtre*" (Earle 1888:190: "*sihtre*" also) and BCS 963 (S 622): "*seohtres*". See EPNE (2:119); ERN (365); PNBrk (787); PNSx (95) and PNWo (392).

*****seoluc** (OE) Not too strong a contender in the watercourse hierarchy. KCD (673) gives the form: "*sioluc*" and BCS (780) has: "*seolciug fleot*". Wallenburg (257-8) discusses the BCS charter and EPNE (2:119) give a list of possible PN's.

server (E) The OED1 (8:514c) gives: "rilles and servers of waters into euery street", which comes from Camden's *Britannia*. It is defined as: "A conduit or pipe for conveying water". In Cheshire, server features in a number of FN's, see PNCh (5:1:1ii:336; 2:201; 3:226 and 4:144 & 296).

seu (E) See sew.

seuch (Sc.) See sheugh.

seuera (L) See sewera.

seugh (E & Sc.) See sheugh and sough.

sew (E) An agricultural drainage term. Wright (841a) gives: "A covered drain or wet ditch" and "sue, a drain." Parish (117) states: "sue, to drain land, also a drain." See the MED, under seu.

sewer (E) From the A-N *sewere*; OF *sewiere*, 'drain or channel'; used in the sense 'a fresh water channel with a bank used for drainage purposes', and not in its modern sense as 'a channel, usually underground, for discharging waste water and refuse from houses'. Callis (99) states: "the Sewer is a fresh water trench compassed in on both sides with a bank, and is a small current or little river." Further, (325) he states: "That water is the substantive of all these: and if it be a running water at random, then it is stream; if it be a running water, and pent within walls or banks, then it is a river, gutter, ditch, or sewer." The term is commonly found on OSX 123. See sewera and the MED, under seuer.

sewera (L) Martin (318b) gives: "sewera A trench to preserve land from floods; a sewer", and (318a) the variant "seuera". Black (96a) gives: "ASSEWIARE. To draw or drain water from marsh grounds". See sewer.

seyke (E) See sike.

sgriob (SGael.) Basically a sgriob is a 'scatch, mark line or furrow'; another case of a word becoming a watercourse generic in a transferred sense. A sgriob is shown on OSX 447 at NC5850.

shadel (E) Wright (842b) gives "shadel, a watergate." See shedele.

shares (E) Axon (160) defines shares as: "rills or streams of water." Axon's work highlights words found in Bailey's *Universal Etymological English Dictionary*, but, from the entry, it would appear that they are his own interpretation of some of Bailey's entries. Under "*SHERBURN Lane*," Bailey refers to the fact that it: "was so called on account of a long *Bourn*, or stream of sweet water ... where running fourth, and breaking into many small rills or streams, it left the name *Sharebourn Lane*."

shedele (E) Wright (846a) gives "A channel of water." The PP (196b) under gote, gives: "water schedellys". See shadel.

sheep-drain (E & Sc.) An agricultural drainage term. Stephens (1848:8) states: "When the grass is smooth and the soil pretty deep, this is an economical mode of making an open *sheep-drain*." He gives a plan of one on p 7 and a sketch on p 9. For the Sc., The DSL (SND) use the same source as the Stephens book was published in Edinburgh and London. See EDD (5:373b).

sheth? (E) Brockett (31) "Sheth, a portion of a field, which is divided so as to drain off the water by the direction of the ploughings, called sheths; ...."

shetlake (E) "The shetlake that rin'th out to-day, Can grind no grist ta-marra", from Verse 60 in *Jim and Nell: A dramatic Poem in the Dialect of North Devon*, published in 1867. The term is defined by The EDD(5:379a) as 'The stream which feeds a mill' and Skeat (1896:64), in his glossary on the subject, gives: "Shetlake, a stream which feeds a shoot."

sheuch (Sc.) See sheugh.

sheugh (E & Sc.) A northern and Sc. variant of sough. Heslop (619) gives: "SEUGH, SHEUGH, SOUGH, a small stream or open gutter running through land. A surface drain." Brockett (188) uses: "seugh, a wet ditch" and Grose (137) gives: Seugh, or Sough, a wet ditch." Warrack has a number of Sc. forms and definitions – (479b) "Schugh, a drain, a furrow"; (496a) "Seuch, Seugh, a ditch, drain, open gutter"; (502b) "Sheuch, Sheugh, a ditch, drain; a furrow; trench; a small stream" and (507a) "Shough, re. a drain; a furrow." See sough and for other variants, The EDD(5:379); The DSL (DOST), for the OSc., form "souch", who also list "huche" as an error for seuch and The (SND).

shoot (Co. & E) For the Co., Thomas (121) renders this term: "A stream of water." Courtney & Couch have two entries (51 & 100) and give, for the former: "West Co. shoot, water led to a point by a pipe or drain, and then bursting out. In Cornwall they often took the place of pumps" and, in the latter: "East Co. shute, a conduit, or fountain of falling water." Jago (1882:263) gives much the same: "A channel of wood or iron for conveying a small stream of water. Also, the watering place where the women fill their pitchers from the "shute." Also, a small stream of water running from a shute or channel." For the E, Young (1799:275) gives: "... the catch-water drain runs all winter, taking the shoot from an extensive range of hills, and bringing in floods much of the finer and richer particles…." Smiles (1864:54) refers to a "leaded shute" used as a distribution channel for water to supply allotted portions to each street in the town of Tiverton, Dev. In other words a conduit, tapped at various intervals, for public water supply. Overall, this term is obviously used of a man-made open conduit, carrying a stream of water, for public supply, but, nonetheless, can still be designated an artificial watercourse.

shore (E) Not normally defined as a watercourse, moreover, a term to be associated with common sewers for foul water discharge, but it is given in a wider sense by Marshall (1796a:63) who states: "shore, ditch or surface drain" and The EDD(5:404a) which gives: "A mountain stream in Lancs."

shot (E & Sc.) This term (as shots) is shown as a watercourse on OSOL 30 at NY9005, but here, the topography suggests a possible variant form of shoot or shoots as there are half a dozen headstreams emanating from the hillside. However, Warrack (506b), for the Sc., gives: "the spout that carries water to a mill-wheel", which is confirmed by Jamieson (4:209b) and The DSL (SND). See trow.

shough (Sc.) See sheugh.

shoulder-drain (E) One of the many agricultural drainage terms given by Stephens (1848:133).

shouldered turf-drain (E) See turf-drain.

shute (Co. & E) See shoot.

síc (OE) See sike.

sic(e) (E) See sike.

*sicel (OE) A dim. of sic, 'very small stream'. See PNDb (17, 199, 695 &749); PNGl (4:171) and EPNS (1:22).

sichet (E, ME & OSc.) See sicket; DOST and the MED.

sichetum (L) The L form of the E *sicket*. Martin (318b) gives: "little stream of water", and (319a), the variant: "sikettus". Bailey mentions these terms defined as: "a small current of water … dry in the summer." The foregoing suggests that these terms are dims., the same as the E corresponding terms. See sicket.

sichter (OE) See seohtre.

sicket (E) A dim. of sike, 'a small brook or rivulet'. Heslop (639) gives this and two other forms – siket and secket. Eyton (187) quotes a deed from 1260/80 which mentions a sichet. See sichetum and the MED, under sichet.

side-cut (E) An agricultural drainage term much the same as side-drain, usually taken off, apparently, from the main drain. Johnstone (64) seems to be the only drainage authority to mention it, stating that: "In some cases, however, it may be necessary to have a few side-cuts from the main drain."

side-drain (E) One of the many agricultural drainage terms given by Ogilvie (2:96c).

side-stream (E) A term used of a tributary by Goodall (55, 75 & 163), here, one belonging to the R Calder.

side-thoroughs (E) A drainage term given by Ellis (3:37).

sihter (OE) See seohtre.

sihtra(n) (OE) See seohtre.

sihtre (OE) See seohtre.

sik (ON) See sike.

*sīka- (Gm.) See sike.

*sīkam (Gm.) See sike.

sike (E & Sc.) From the ON *sik* and OE *síc*, B-T (869), also found in Ic., The GM. forms are **sīka-* and **sīkam* and, the IG is *seik*. WW (1:652.33) give: "*Hic riuus*, Ae syke" from the 15th C *English Vocabulary*. Pegge (62) gives: "sick. a brook when very small." Wright (855a) has two definitions: "SICE, a gutter, or drain. Somerset", and "SICK, a small stream, or rill." There are no sik- or syk- forms in Wright. Leigh (179) has: "Seech, Seek, Sike or Syke" and Addy (214) also lists sick. Dickinson (327) gives: "Syke", and "Seyke, a water course frequently dry in summer." Three forms of the term can still be found on maps in Eng. – sick on OSL 119 at SK2968; sike on OSX 307 at NY9948 and syke on OSOL 5 at NY4129. For the Sc., Wyntoun (2:401, line 3912), gives: "For a gret syk betwene thame was" and Warrack (510a) defines sike as: "a small rill; a marshy bottom or hollow with one or more small streams."

Two Sc. forms are shown on maps – sike on OSX 321 at NX9595 and syke on OSX 328 at NS7020. See Field (1972 & 1993 for a number of variants); the IED (532a); the MED, under sich(e), for a massive list of forms, surnames and PN's; The DSL (SND); EPNE (121-2) and the EPNS volumes for Brk; Db; Gl; Nt; We; and YW.

siket (E & OSc.) See sicket and DOST for other forms.

sikettus (L) See sichetum.

sioluc (OE) KCD (673) gives the only instance of this term. DEPN (402b) suggests that Silkmore in Staffs; is rooted to it as does Horovitz (550). See PNE (2:119-120) and PNLei (1:202).

sipe (E & Sc.) According to OED1 (9:97) this term is related to Dut. and Fris. Peacock (186974b) gives: "Sipe, a small trickling stream", confirmed by The EDD(5:446b). In OSc. sipe was used of a watercourse but in ModSc. more as a"trickling stream". See The DSL (DOST & SND).

sirkel (SNn.) Jacobsen (1928:758a & b) gives: "a narrow, open drain in the ground, a ditch."

sitch (E) The southern form of the northern 'sike'. Addy (216) gives: "a dyke, ditch, or ravine." and adds: "Near Horsley Gate, Dronfield, are Bole Hill Stick and Salter Sitch. Each is at the bottom of a valley watered by a stream. Sitch = sick … Ditch. 'Sytche, a ditch.' Holland's Cheshire Glossary. The word is commonly used in this district." As Addy has mentioned, Holland (350) gives the form: "SYTCHE. a ditch." A term shown on OSOL 1 at SK1483 and OSX 245 at SK1215. See sike; PNDb (17) for a stream name and PNSa (6:22) for a PN.

siubhlachan (SGael.) Defined by Dwelly (847b) as: "stream, rivulet."

skerth (E) From the ON *skurðr*, 'a drainage channel'. Clarke (291) gives: "delphs and Skerths"; with further mentions on (300 & 303). OSX 301 at TA0331 shows 'scurf' and 261 at TF1755, 'skirth'. See EDD (5:481); EPNE (2:126); PNYW (4:15) and PNWe (2:74).

skeugh (E) From the ON *skógr*, not normally defined as a watercourse but is shown as such on OSOL 30 at NY8900. See

Atkinson (1868:457); EPNE (125-6) and the PN references quoted therein.

skirth (E) See skerth.

skógr ON) See skeugh.

skor (ON) See score.

skurðr ON) See skerth.

slaak (Fl.) See slack.

slack (E & Sc.) From the ON *slaaki*. The validity of this term looks questionable; most dictionaries define slack only as 'a small shallow dell or valley'. Halliwell (752b) gives "SLAKE, a deep ditch; a ravine", which is shown on OSL 75 at NT9852, as a watercourse. Other slacks are shown on OSOL 26 at NZ6012 & 4604. For the Sc., Jamieson (278a) gives: "SLAK, Slack, Slake, 1. An opening in the higher part of the same hill or mountain, where it becomes less steep, and forms a sort of pass" and (278b), "Slack, a valley or small shallow dell". Sc. slacks are shown on OSX 336 at NT1729 and OSX 427 at NJ9941 and other slacks are shown at NGR NJ3758 and nearby. See EPNE (2:128) and compare Fle. slaak, 'stream' in USBG (Bel.:ivb).

slade (E, Ir. & Sc.) Wright (1880:865a) describes slade as "a dried watercourse (Essex)". Cope (83) gives: "a brook; a small running stream", a term used in Hampshire. Noteworthy too, is Joyce (1869:2:387), who brings our attention to the fact that: "In some eastern counties of Ireland – especially in Dublin – they apply the word slád or slade to a stream running in a mountain valley or between two hills." Jamieson (4:274b) has a quotation which states: "a hollow between rising grounds, especially one that has a rivulet of water running through it", and Warrack has the same. However, it would appear that the only mapped courses are to be found in Essex. There are three marked on maps, two on OSX 195 at NGR TL5335/8 and another on OSX 174 at TQ4499. The first two are mentioned in the PNEss (7, 11 & 16, under The Sledway) the other, Genesis Slade, does not appear to have been listed. In ME, slade is sometimes, but not exclusively, given in the form 'slæd', on this, see the MED under slade.

slæd (ME) See slade.

slak (Sc.) See slack.

slake (E) See slack.

slakki (ON) See slack.

sloc (SGael.) Dwelly (855a) gives: "hollow, cavity, pool, gutter, ditch." the extended form slochd shows two examples on OSX 390 at NM4771 and 4871 and Darton (251-2) covers the PN forms. Middendorff (118 under slohter) mentions the Nth. word slochter which corresponds with the Latin fossa – 'ditch', and, which may be relevant here. The only current G word to come near is schlucht, defined as 'gorge, ravine'.

slochd (SGael.) See sloc.

slog (Sc. & SNn.) Milne (293) gives: "The Slogs is the valley of a very small stream." Jacobsen (833a & b) gives: "sloggi, a ravine, a sunken path ... sloping mill-race ...." and roots it to the "Nor. slok", which suggests a possible connection with the ON *slakki*, although the SGael., sloc may have a claim too as sloc has meanings such as 'hollow, cavity, pool, gutter, ditch." A mapped example is shown at NGR NJ4322.

*slohtre (OE) The form in AD (556) is: "slohter" which refers to Slaughter in Gloucs. Slaughterford is given as: "*slohtran ford*" in BCS (230) and "*slohterword*" in (882). See EPNE (2:130) and the PNGl (1:206) for the watercourse terms associated with G and LG.

slonk (Sc.) Warrack (531a) gives: "Slonk, n. a mire; a ditch; the noise made by wading or sinking in a miry bog, and when walking with shoes full of water ... to wade through a mire; to sink in mud." The DSL (DOST & SND) only define slonk and slunk as 'hollow, ditch'. A slunk is shown on OSX330 at NT2416.

sloped-gaw (Sc.) An agricultural drainage term given by Warrack (531b) who states: "an open drain", Jamieson (293a) gives the same. See The DSL (SND) under gaw.

slough (E & Sc.) From the OE *sloh*, hardly any references exist for this term in word books, the OED1 (9:236c) has a few –

EDD – none, in the sense 'watercourse'. Sc. dictionaries generally define this term as 'marsh, quagmire or bog'. However a Sc. slough can be seen on OSX 345 at NT6269.

slugach (SGael.) See slugaid.

slugaid (SGael.) Dwelly (856b) defines this term as: "slough, deep, miry place." A slugaid is shown on OSX 460 at NB5462 and its variant slugach on OSX 453 at NF7515.

sluice (E) 'A channel cut for the purpose of carrying off surplus water'. Leland (1:142) states: "Ther goith a sluse out of this bath, and servid in tymes past with water derivid out of it." Two sluices are shown on OSX 285 at SD4215 and another at NGR SD4116.

slunk (Sc.) See slonk.

sluse (E) See sluice.

small main (E) Dickson (430) gives: "Small Mains. These are the next gradations of ditches applied to distribution, by means of their communication between the head main and trenches."

snuadh (Ir. & SGael.) Windisch (784b) gives: "river, brook". O'Brien, O'Reilly and Lhyud all confirm – Lhuyd's entry (411c) compounds the term with chlais: "snuadhchlais, the channel of a river." For the SGael., Dwelly (865b) defines this term as: "obsolete for river, brook."

snuadhchlais (Ir.) See snuadh.

soak ditch (E) Artizan (128a) gives: "... at the point where the bottom of each trench comes to coincide with the soak ditch at the top of the slope, the water contained in that trench will be drained into the open soak ditch, and carried off, according to the natural fall of the ground."

soak-drain (E) A soak-drain is shown at NGR TF0099.

soak-dyke (E) See sock-dike.

soch (Wel.) The Wel. form of the E sock. Pughe & Pryse (2:523a) give: "a drain, a channel for water." See sock, sock drain and sock dyke.

sock (E) Rye (204-5) states: "(3) The mouth or outlet from a ditch into the river. (4) More generally and correctly the ditch running parallel with the river outside the wale." See soch, sock drain and sock dike.

sockage drain (E) Young (1799:284) mentions sockage drain as: "a drainage and warping term."

sock-dike (E) Wright (882b) gives: "A ditch on the inside of a marsh embankment to carry off the water which soaks through it. Norf." Peacock (1889:505) gives: "SOAK-DYKE, SOCK-DYKE.– A ditch beside a large drain or canal, for the purpose of receiving the water which percolates through the bank." A soak dike is shown at NGR TF5605 and a soke dyke at TF2320.

sock-drain (E) A compound drainage term shown at NGR TL4671.

sock-dyke (E) See sock-dike.

sod-drain (E) An agricultural drainage term given by Stephens (1848:133) who states: "An imperfect form of shoulder-draining is practised in some parts of England on strong clay soils, under the name of sod-draining." and further that: "The sod drain is very similar in construction to the covered sheep drain in grass".

soke-dyke (E) See sock-dike.

souch (OSc.) A variant of seuche. See sheugh.

sough (E, ME & Sc.) A term used mainly in the north and Scot. Holland (345) and Peacock (535) both give: "suff"; Wright (822b) quotes Drayton (1:99) who gives: "Saugh", and Loudon (1247) defines it: "a box-drain." Heslop (619) gives three forms: "SEUGH, SHEUGH and SOUGH, a small stream or open gutter running through land. A surface drain." Many other word books list this term in the sense 'drain' and Hartshorne (*Salopia Antiqua*, 581) was of the opinion that suff was: "entirely our own", whereas we know that it is widespread throughout the North. OSX 314 at NY2352 and OSOL 4 at NY0425 give mapped examples. For the Sc. Warrack (548b) gives: "Sough, a ditch; a trench." See sheugh; PNLa. (75) for one of the few PN connections and the MED.

sow(e) (E) A variant of sough listed by Carr (2:150); Easther (125) and Dickinson (356). Prevost (156a) gives the variant sowe. See ERN (375) for the RN's

spinney? (E) Wright (895b) gives: "A brook. Bucks." Halliwell (784b), under spinney, cites a line from *Sir Gawain and the Green Knight*: "At the last bi a littel dich he lepez over a spenne". According to Madden (412a) a *spenne* is a hedge, so an error by Wright, perhaps, for thinking that it meant ditch rather than hedge.

spong-water (E) Wright (897b) gives: "A small stream" and Forby (2:320) states that: "Spong-water is a narrow streamlet."

spout (E & Sc.) Blakeborough (138) gives: A small stream of water filling a trough, a stream from a natural spring. This term is more commonly found in Scot. Warrack (557b) gives: "a runnel of water." OSOL 32 at NX4682 and OSX 330 at NT1215 are two examples and the term is also used of a waterfall.

spray-drain (E) An agricultural drainage term. Ogilvie (4:170b) gives: "a drain formed by burying the spray of trees in the earth, which serves to keep open a channel. Drains of this sort are much in use in grass lands."

spring (E) 'A rising or issuing of water from the ground'. Well represented on maps – shown at NGR SD7061; SD7776 and 'springs' at SD7373. See the MED.

spring-drain (E) An agricultural drainage term. Rees (33) gives: "that sort of drain or channel which is prepared and constructed in land for the purpose of taking away such kinds of over-wetness in it ...."

sput (Ir. & SGael.) Related to spout – O'Reilly (487a) gives: "an eunuch; a spout, an aqueduct." There is, in Scotland, a watercourse bearing the name Sput Ban which can be seen on OSX 364 at NN3123 and its dim., sputan, is shown on OSX 414 at NH1121 and OSX 417 at NH6630, however, in SGael., the sput term is normally defined as 'cascade, waterfall' rather than a generic term for a watercourse. See spout.

sputan (SGael.) See sput

srabh (Ir.) O'Reilly (487a: srabh) and, in particular, (703b: sraobh) gives a very interesting story relating to this word: "a stream:

hence Shrove or Sreeve Point, in the parish of Lower Moville, barony of Inishowen, county of Donegal. A trickling or vein of water oozes from a rock here, to which all the deranged people of the country are wont to resort." OSIDS 23 at F8511 and & 31 at F 9906 & 9930 show srah as RN prefixes, a possible contraction of srabh, although srah is normally used of a river holm it would appear that, here, at least, the term may have become a transfered RN but must be left open to question pending further research. See srubh.

srae (Ir.) According to Joyce (1869:2:221-2) srae and sraobh is used to denote "a mill stream." See srabh.

srah (Ir.) See srabh.

sraobh (Ir.) See srabh and srae.

sreamh (Ir.) O'Reilly (488a) and (488b, sreimh) gives these forms for denoting: "a stream or rill."

sreb (Ir.) This can be found in the ALI (1:26 line 27), although sreb is the base form there are many variants – for a full list see eDIL.

sreban (Ir.) The dim. of sreb, denoting a streamlet, found in Earls (106:21).

sreimh (Ir.) See sreamh.

*sreumen (IG) See stream.

*srew (PIE) See sruaimh.

srib (Ir.) Windisch (791b) furnishes us with this form of sreb: "a stream." See sreb.

sroth (Ir. & SGael.) For the Ir., Lhyud (413c), O'Reilly (489a) and O'Brien (412b) all give: "stream, brook" or the like – in SGael., sroth "stream" is given by Dwelly (894b).

srothan (Ir.) The dim. of sroth, 'streamlet' – listed by O'Reilly (489a) and O'Brien (412b).

*srowman (PC) See sruaimh.

sru (Ir.) The sru- stem has produced a multitude of forms – there are twenty listed below – Joyce (1869:3:338 & 562) gives it as a contracted form of sruth, which he says survives in Drumsru, Kildare. See all the sru- forms.

srua(i)m (Ir.) See sruaimh.

sruaimh (Ir.) From the PC **srowman* and PIE **srew*, given as OIr., by the majority, O'Reilly (489a) gives two forms – "sruaimh and sruam". Windisch (792a) and eDIL use sruaim, variants of each other meaning 'stream'. Cormac (153, under sron and sruth) gives: "sruaim". Pokorny (1003) gives: "sruaimm".

sruaimm (Ir.) See sruaimh.

srubh (Ir.) Joyce (1869:3:562) gives this term and its variant sruibh with PN's examples – Sriff in Leitrim and Sroove in Galway – both mean 'stream'. Stroove and its Gaelic form 'an tsruibh', in County Donegal, is shown on OSNIDS 4 (Coleraine) at C6642. See srabh and Moville in the bibliography, as the derivation varies.

sruffan (AIr.) Joyce (1:458) gives this as an anglicization of sruthan.

sruffaun (AIr.) This is an old form of sruthan, an anglicization of it. Sruffaunglass shown on the 6" OIM, is Sruthan Glass on later maps. There are other sruffaun- stream names shown on OSIDS 45 - Sruffaunbeg at M1528, Sruffaunboy M1827 and Sruffaunree M1024, a small stream flowing into the ocean at Spiddle. See sruthan and Joyce (1869:1:458 & 3:536).

sruh (Ir.) Joyce (1869:3:562) states: "sru and sruh represent the Irish *sruth*, a stream." Both, therefore, are contracted variants of it. There is a sruh prefixed RN shown on OSINDS 26 at H1032. See sru and sruth.

sruhan (Ir.) Used as a prefix to stream names on some NI maps such as OSNIDS 12 which gives four examples; also shown on OSNIDS 7 at C6500. This is a contracted variant of Sruthan – the pronunciation of it. Joyce gives (1869:3:562) some PN examples, one of which has a rare double 'a' form: "Sruhane in Tipperary, and Sruhaan in Leitrim and Wicklow." See sruthan.

sruhaun (Ir.) See sruthan.

sruibh (Ir.) See srubh.

sruill (Ir.) This term is listed in eDIL and means 'stream'. Sroohill, discussed by Joyce (1869:1:48) is one of the few PN's together with Shrule and Shruel which root back to a variant of sruthair; Sroolane in Limerick is probably another.

sruth (Ir. & SGael.) From the PC **sruto*. The most common term for a stream used in Ireland and the Gaelic regions of Scotland (not found in Mx., here, the forms are strooan and struan). Listed in all dictionaries, eDIL in particular, it is also used extensively as a PN element. Joyce (1869:1:457) has numerous examples and variants – Srue in Galway; Shruh in Waterford and Shrough in Tipperary, to name just a few. For the SGael, there is an entry in Dwelly (895a) as well as map evidence – one can be seen on OSX 357 at NR7663 and CPNS has many forms, all listed in Index (E). See sruh.

sruthail (Ir. & SGael.) The term survives in Ballyshrule (Irish – Baile Sruthail), 'Townland of the stream', in Galway. Joyce (1869:3:6) states: "for it is locally pronounced by the best authorities Baile-sruthra". For the SGael, CPNS (503) would appear to be the only reference for this watercourse dim., further evidence is lacking as most dictionaries give it in a verb sense.

sruthain (Ir.) See sruthan.

sruthair (Ir. & SGael.) Another dim. of sruth, meaning 'streamlet, brook, rill, rivulet'. Joyce (1869:1:457-8) gives PN examples stating that: "Abbeyshrule in Longford was anciently called sruthair, i.e. the stream." Windisch (792b) gives sruthar and sruthair is listed in eDIL and CPNE (503).

sruthan (Ir. & SGael.) Another dim. of sruth, 'streamlet, rivulet'. A NI instance is shown on OSNIDS 13. O'Reilly (704a) gives the form sruthain, most others – sruthan. Sruthan Glass in Galway is shown on OSIDS 45 at M1630 which, on OIM, is sruffaunglass! Further references can be found in eDIL and Joyce (1869:3:143) lists the PN Boleynasruhaun, Galway. Other PN prefixed variants of sruthan, given by Joyce (1869:3:565), are: "Stroan and Struaun, in Mayo", the latter is shown on OSIDS 31 at M1189 and OSIDS 44 at L7533 shows another. The SGael. is covered by Dwelly (895b) and CPNS has various entries listed in Index (E). A Scottish sruthan can be seen on OSX 355 at NR4777. See sruffaun and sruhan.

sruthanan (SGael.) A dim. of sruthan, "streamlet, brook, rill, rivulet", given by Dwelly (895b) and shown on OSX 355 at NR4266.

sruthar (Ir.) See sruthair.

sruthchlais (Ir. & SGael.) For the Irish O'Reilly (489b) gives "the channel of a brook." The same definition is given by Lhuyd (413c). For the SGael., Dwelly (895b) gives: "Water-channel, conduit, canal. 2 Bed of a river or stream."

sruthlag (SGael.) Dwelly (895b) gives: "small brook, rill". Maps show a number of 'lags' but no 'sruthlags'. See lag, log and sruthlog.

sruthlog (Ir.) Coneys (342a) defines this compound as: "a rivulet, a rill, water conducted through a pipe ...." A term confirmed by Dineen (1904:689a). See sruthlag.

*sruto (PC) See sruth.

staing (SGael.) Dwelly (897a, staing & 898b, the variant stang) gives: "ditch, trench." Two instances of staing can be found on the modern map, one on OSX 359 at NM8108 and another on OSX 360 at NN0425.

stall-drain (E) An agricultural drainage term. Dickson (1807:1:76) gives: "main drain, into which all the stall-drains should empty themselves."

stang (E & OSc.) See stank.

stang (SGael.) See staing.

stank (E, OSc. & Sc.) From the F *étang*, 'pond, pool'. According to the OED1 (9:823) the earliest reference to stank, and its variant stang, comes from the *Cursor Mundi*, dated 1300. Wright (906b) gives: "a wet ditch." For the Sc., Warrack (564b) gives: "a stagnant or slow-flowing ditch." The term is also mentioned by Wyntoun (2: line 4407) "By a stank at a gate syde." The term stanks is shown on OSOL 42 at NY8085 and the variant stangs on OSOL 4 at NY1244. A Sc. stank is shown at NGR NK0532. See The EDD (5:729b) and The DSL (DOST & SND).

stell(e) (E & Sc.) Heslop (690) gives: "STELL, STILL, the water channel running through a marsh...." Dickinson (314) and Atkinson (1868:495) list the term also, and Brockett (159) adds the form: "Stelle". A stell and main stell (one of the longest) are shown on OSOL 26 at NZ5513 and a Sc. stell is shown on OSX 336 NT0326. See water-stell.

ster (Bret.) PNWe (1:15) suggests the possibility that the etymology of R Winster might have some relativity to a line in the XI *Book of Taliesin*, "Kat y gwensteri ac estygi lloygyr", 'A battle in Gwensteri, and thou subduest Lloegyr'. Gwensteri, is given by a number of authorities as Wel., gwen – 'white' and the Bret. word, ster – 'river, stream', but contemporary opinions still differ about the Winster etymology. In Brittany, ster and steir are used as the generic appellation to RN's. See avon.

stickle (E) This word is used of running water but not normally as a generic term for stream. The glossary entry given in Skeat (1896:64b), following the *Jim An' Nell* poem, would appear to be the only instance of stickle defined as "a small stream".

still (E) See stell and water-still.

stitch (E) Morris (1857:63) gives: "Stitches (Anglo-Saxon), deep narrow furrows for draining land." but, the usual definition of stitch is a ridge between water-furrows and not normally the watercourse itself.

stone drain (E) An agricultural drainage term. Arkell (325) gives it in his paper *On the Drainage of Land*, stating that: "Stone drains are various; the most common here are wall and dribble, or rubble, the former as main, the latter as tributary." Stephens also covers stone drain, together with sketches, in *The Book of the Farm* (1889:5.263b) and, in addition (1848:124-6), discusses Stone drains at length under the heading "Flat Stone Drains." See the alternative names – dribble, rubble and wall drain.

stone-gutter (E) An agricultural drainage term. Elworthy (573), under pipe-gutter, gives: "... in distinction from a stone-gutter, which is one made of loose stones, until late years by far the commoner kind."

stoop (E) See over-shot.

stræm (ME) See stream.

straits? (E) Although OSOL 43 at NY9855 shows Boghall

Straits as the upper reaches of Boghall Burn, straits is normally defined as 'a narrow passage for water connecting two seas', so it must be considered questionable.

stram (ME & OFris,) See stream.

strame (E) See stream.

stran' (Sc.) See strand.

strand (E, ME, O & ModSc.) Possibly from the OE *strand*, 'shore'. The OED1 (9:1076) gives: "A stream, brook, rivulet." and suggests a connection with strind. Wycliffe (3:270b, from Isaiah 27:12) and (4:286, from John 18:1) gives: "stronde" and "strond" respectively. For the Sc., Warrack (577b) gives: "Strand, Stran', a stream, rivulet; a gutter; a channel or drain for water." More references are given in The DSL (DOST & SND). At least five strands are shown on OSOL 16 and a Sc. one can be seen on OSX 328 at NS5703. See strind; strine; The EDD(5:804a); the MED, under stronde and PNWe (2:291).

*strauma (Gm.) See stream.

*straumaz (Gm.) See stream.

straumr (ON) See stream.

straun (Sc.) Warrack (578a) gives: "straun, a rivulet; a gutter" and "strawn, a stream; a gutter." Jamieson (4:438b) under strawn, gives: "a gutter". See The EDD(5:804a) under strand.

straw-drain (E) One of the many agricultural drainage terms given by Ogilvie (222a): "a drain filled with straw." Rees (34) also mentions the term.

strawn (Sc.) See straun.

stream (E, Co., Ir., Mx., ME, NIr., OE, OSc., Sc. & Wel.) From the Skt. *sru*, 'stream, flow'; 'water flowing in a natural channel, such as a river or brook'. There are many cognates, the majority of which are listed in Köbler's dictionaries – Dan., *strøm*; Dut. and LG *stroom*; OFris., *stram*; G and OHG *strom*; OHG, *stroum* and *strum*; ON, *straumr* and OSax., *strom*. In addition to the Gm. list, the MWel., frut and frutt; ModWel., ffrwd and all the Ir. and Sgael. sru- and stru- stems as well as the Mx. stro- stems can all be added here. The Gm. forms are **strauma-* and **straumaz*, and the *IG* is **sreumen-* (Köbler) and

(Pokorny, 1003). For the E, Salibury (60) gives: "strem", a ME form and Lously (156) gives: "STRAME OR STREAM", the former, almost a match for the OFris. stram! In Co. strem is used as a generic, which George (2009:604b) gives as one of the: "partly assimilated loans"; see 5.2.3 Styles of print (20). Nance (1978:295b) gives: "stream". Ir. dictionaries do not give stream; it only appears on maps, through the anglicization of the indigenous terms. Layamon gives a number of ME forms – stræm in (2:405); stram in (2:469) and streme in (1:261). For the OE, B-T (926) give: "*A stream, current, flowing water*" and the OSc. terms are given under strem(e) in The DSL (DOST). Kelly, Kneen and Moore cover the Mx. terms under the stroo-stem, and like the Ir., the ModSc. and Wel. terms only appear on maps. For an E mapped example see OSX 144 at SU7737; Ir. on OSIDS 64, which shows at least 8 streams; NIr., on OSNIDS 29 at J2528, 2625 and 3824; Sc. on OSX 336 at NT1334; Mx. on IOMS at SC3874 and Wel. on OSX 255 at SJ2643. The (Da.) strøm, USBG (Den.:viia); (Fl.) stroom, USBG (Bel.:ivb) and (G) Strom, USBG (GDR:xvii) terms are still currently in use. Buck (41:1.36) has probably one of the most compact and comprehensive lists published, covering the IE cognates, and, for a list of Indo-Aryan ones, see Turner (1962:802-3). The MED has a substantial list of quotations and variants, and many others are listed in the editions of The OED. In addition, all the stems listed above should be consulted for a full digest, and, for associated PN's, see PNBrk (3:788-9 & 908); PNGl (4:176); PNNt (267) and PNYW (7:252).

streamie (Sc.) Warrack (578b) gives: "Streamie, a streamlet."

streamlet (E) The dim. of stream, Leland (2:37) and Holinshed (1:112) both give the variant "streamelet".

streamlic (OE) Clarke Hall (278b) gives: "*streamlic of water.*" See *licc.

streamrace (OE) WW (1:345.22) give: "*aleum*, streamrace." from an 11th C *Latin and Anglo-Saxon Glossary*.

streám-racu (OE) WW (1:178.5) give: "*streamracu*, alueus" from

The Supplement to Alfric's Vocabulary and B-T (927) states: "*The bed or channel of a stream, a water-course.*" See rake.

streám-rád (OE) B-T (927) give: "*the bed, course of a stream*."

streamum (L) See torrentibus.

streamway (E) The OED1 (9:1098b) gives: "the main current of a river" and "the shallow bed of a stream."

streap (Sc.) Warrack (578b) gives: "Streap, a rivulet." and (580a) "Stripe, Strip, a rill; a small stream; a small, open drain; a long, narrow plantation", and, The DSL (DOST & SND) have a good list of variant forms under strip(e). A stripe is shown on OSX420 at NJ3530 and a strype on OSX 420 NJ3928.

stredh (Co.) This word has a few of variants – stret(h) and streyth. For stredh, George (1998:132b & 2009:604b) gives: "stream, brook." Williams (1865:326a) lists stret and gives: "a fresh spring, a stream" and further states that this word is: "written also streth, or streyth." Nance (1978:296a) also lists streth, for which he gives: "stream, brook" and, in addition to Williams, streyth can be found in Norris (422-3) whose reference comes from the *Ordinale de Origine Mundi*, line 772. See CPNE (213).

strek (Co.) In the works of Borlase , Pryce and Jago (1887) this term is listed as "stream." As with many of the entries the Jago listings tend to follow earlier authorities.

strem (Co & ME) See stream.

streme (ME) See stream.

strengr (ON) See string.

stret (Co.) See stredh.

streth (Co.) See stredh.

streyth (Co.) See stredh.

strin(n) (Sc.) Warrack (579b) gives: "Strin, Strinn, a thin, narrow stream of water, &c.; the channel of a river ...." Gregor (185) gives strinn also, defined as: "to flow in a thin, narrow stream." See The EDD (5:820a) and strintle.

strind (ME, O & ModSc.) The OED1 (9:1138a) gives: "strunde", "strynde", "strind" and the dialect form: "strine. A stream, rivulet." Stratmann (585b) also gives strunde. For the Sc.,

Jamieson (4:447b) gives: "Strynd, stream, rivulet" and states further that: "Strynde occurs in old deeds, as denoting the course of a rill." The DSL (DOST & SND) also list strind. The PNCu (323); PNDb (3:696 & 751); PNWe (2:291) and PNYW (7:253) cover strind as does the MED under strind(e) and strond(e). See strand and strine.

strine (E) According to the OED1 (9:1138a) strine is a dialect form of strind. Addy (57) gives: "a ditch." and Jackson (417): "a water channel." Loudon (1146) states: "These brooks are known in the country by the name of Strines, being distinguished from each other by the name of the places from which, or past which, they flow." This term is fairly common in R and PN's: strine (pipe & brook) is shown on OSL 127 at SJ6918; Strine (river) on OSL 127 at SJ6518; Strine Dale (PN) at NGR SD9507; Strines (PN) at SJ9786 and another Strines (PN) at SK2290. On Dartmoor there is a R Strane, shown at NGR SX6171, the etymology of which is obscure, so it may belong here. See strand; strind; Horovitz (2:360 & 576 for variant forms); The EDD (5:818b); EPNE (2:163, under strind); PNCh (1:65, 268 & 5:1.ii: 357); PNDb (1:152 & 3:606); PNSa (5:215) and, for a number of mentions in Yorks., PNYW (1:228 & 327; 3:88, 186 & 199).

string (E) From the ON *strengr*, 'watercourse'. The pluralized form 'strings' is shown on OSOL 30 at SD9994. See EPNE (2:163); PNC (346) and PNYN (231).

strinnle (Sc.) See strintle.

strintle (Sc.) Warrack (580a) gives: "Strintle, Strinnle ... to flow in a small stream ... a very small stream...." The DSL (SND) list strinnle, under strind, as a dim. form and Gregor (185) also gives: "a very small stream." See strin(n)

strip(e) (Sc.) See streap.

stripie (Sc.) Warrack (581b) gives: "Strypie, a very small rill", and Jamieson (4:443b): "Stripie, Strypie", the same. See streap.

strippet (Sc.) A dim. of strip, Ogivie (4:227c) gives: "A small brook; a rivulet." See streap and stripie.

strom (G, OHG & OSax.) See stream.

strøm (Dan.) See stream.

strome (A-N & E) The E term looks like ON *straumr*, 'stream', as suggested in PNNt (8) rather than A-N "strome, stream" listed by Kelham (228b). MacBain (1922:56) applies the same ON root to Stromeferry in Scotland. A 'strome' can be seen on OSX 300 at SE5965, its geographical location being within the bounds of N survivals. See Köbler (ON) and Pokorny (1003).

strond(e) (ME) See rinnon and strand.

strone (Sc.) The EDD (5:824b) gives this, defined as "A stream; a runlet of water" and "strune", a variant, from a quotation therein. See strown.

stroo (Mx.) Kelly (177a) gives: "stroo, a stream; strooan, a stream of water", and "stroo-ny-hawin, a river or brook." In addition, (221a) he gives: "strooanolt, brook." Moore's PN book (158) gives: "STROAN-NY-CRAUE, 'Stream of the Bone'." and further states that: "There is nothing in Manx to correspond to the Irish and Gaelic sruth." Other stream names are given on p 170. Kneen has quite a few stream names too, in particular, under Struggan Snail, he states that: "The first element is a Southside pronunciation of *Strooan,* Ir. *Sruan,* 'a stream'." Another stream name, Struan Barrule is shown on IOMS at SC2877. Johnston (229) and Ross (206) both list a Struan PN in Scot., and there is another Struan on the Isle of Skye. In SGael., sr becomes str in some Scottish PN's and stream names, so too, perhaps, in Mx.

strooan (Mx.) See stroo.

strooanolt (Mx.) See stroo.

stroom (Dut. & LG) See stream.

stroum (OHG) See stream.

strowan (SGael.) Maxwell (216) gives this as a variant of struan.

strown (E) Robinson (1876b:138a) gives: "a runlet of water, answering the purpose of the 'sike,' but not having the same force of current ...." See strone.

struan (Mx.) See stroo.

struhan (Ir.) Another sruhan variant, a rare form shown on OSIDS 73 at R5117.

strule (Ir.) The PN and RN form of sruthair. Kilstrule in Tyrone is shown on OSNIDS 12 at H3485, which also shows the tautological Strule River at H3986.

strunde (ME) See strind.

strune (Sc.) See strone.

struther (Sc.) Milne (undated:46) gives: "Struther. Stream. Sruthair (irish), stream." and (18 & 51) a couple of stream names using this element. Two are shown on OSX 345 at NT5453 and NT5754. For the PN Anstruther, see Ross (10).

strynde (ME & OSc.) See strind.

strype(Sc.) See streap.

strypie (Sc.) See streap and stripie.

stuffing drain (E) An agricultural drainage term. Rees (34) gives: "a term applied in some places to the practice of filling them with wood, straw, or other materials, as is the custom in some instances of surface draining."

sua (Ir.) O'Reilly (493b) defines this as: "a little stream."

sub-drain (E) An agricultural drainage term which can be found in Marshall (1787:264) "... he is making a trench for a sub-drain!"

sub-main drain (E) An agricultural drainage term, given by Stephens (1848:88), obviously a term below main-drain in the drain hierarchy.

succour (E) 'A river tributary' – Holinshed (1:88) gives: "... and meeteth again with a succor of ditchwater ...." Lambarde (199) also mentions the term "... concerning one of the succours to Medway."

sue (E) See sew.

suff (E) See sough.

suir (Ir.) O'Reilly (496a) gives: "water, a river."

sulcus (L & OE) PP (517) gives this under water fore. In ModL it means 'furrow, trench'. For the OE, Bailey gives: "SULCUS *Aquæ* [Old Law] a small Brook or Stream of Water." Black (1122a) gives the same: "In old English law. A small brook or stream of water." The OE sulh, 'plough, furrow' is a possible relative and Kiliani (486a) lists: *sille; sijle* and *sulle*, "aquagium, aquaeductus", which may be relative too. See EPNE (2:167).

sulh (OE) See sulcus; water-furrow; EPNE (2:167); PNBdHu (42) and PNMx (6).

summer-trench (E) See trench-royal.

sunk-pit drain (E) An agricultural drainage term given by Rees (34) "a term sometimes applied to that sort of drain which is sunk large and deep, for the purpose of drawing off the wetness of the land, or for collecting, receiving, and holding the water of it, until it can be removed by other suitable means, small common superficial drains in the land being made to communicate with it, so as to render the ground perfectly dry."

supply (Ir.) An E term shown on an Ir. map; corresponding in the same sense as that of canal feeder and shown on OSIDS 48 at N2225 and 2424.

surface-drain (E) An agricultural drainage term. Loudon (582:3609) gives: "The surface-drains, or water-tables, should be made a few inches lower than the side of the road, and of the common width of a spade at the bottom, and they should have frequent cross drains under the path and fence, back into the outer side drain."

suspiral (E & ME) The OED1 (9:261b) give: "A pipe or passage for water leading to a conduit." See CLB (105).

swamp (E) Not normally defined as a watercourse but is shown as such on OSOL 41 at SD7561.

swash (E) Bailey, Axon (183) and Wright (932b) all give this variously defined as: 'stream, puddle, torrent of water'.

swin (E) Turner (1901:286) gives: "Swin, Swine; a stream." The Fle. word zwi(j)n is defined as "drainage ditch", by the USBG (Bel.:ivb) and is, therefore, very relative here, however, it is not listed by Kiliani but *is* listed in the PNYW (7:287). See PNEss (16); EPNE (2:172); PNYE (51) and PNYW (2:10).

swirril (E) Robinson (1876a:191) gives: "a rill, falling steeply down a hill-side." Heslop (2:713) mentions: "The *Swirle.* a small runner ...", which may be related.

syk (Sc.) See sike.

syke (E & Sc.) See sike.

sytche (E) See sitch.

T

table (Sc.) Warrack (593a) gives: "a watercourse at the side of a road to carry off water."

tabling (E) The EDD (6:2b) gives: "The side of a road or path, having an entablature of soil, along which the water runs; a kind of gutter." See water-table.

tae (Sc.) Warrack (593b & 618b) gives: "Tae" and "Toe" respectively – "... the branch of a drain." Taes are shown on OSOL 31 at NY8137 as watercourse elements.

taidiu (Ir.) This is given by eDIL in the sense "watercourse, leat." Further references are lacking.

tail (E) Peacock (1889:548) gives: "TAIL, TAIL-WATER – The water which has run beneath the wheel of a water-mill." A tail is shown on OSX 145 at TQ0558. The Sc. *tail* is only found compounded with other terms such as –dam, -lead and -race. See the MED.

tail-dam (O & ModSc.) DOST give this as: "the tail-race of a mill"; tail-stream and tail-lead given in The DSL (SND) have the same meaning.

tail-drain (E) Loudon (4414) states: "The tail drain is a receptacle for all the water that runs out of the other drains, not so situated as to empty themselves into the river; and therefore it should run nearly at right angles with the trenches, but, in general it is drawn in the lowest part of the ground, and used to convey the water out of the meadow where there is the greatest descent." Ogilvie (4:297c) mentions the term too.

tail-goit (E) Easther (56) gives: "The channel which conveys the water from a mill is called the tail goit"

tail-lead (Sc.) See tail-dam.

tail of the mill (E) Elworthy (1886:734) gives: "That part of the channel or water-course which conveys the water away from the water-wheel." See mill-tail.

tail-race (E & Sc.) Ogilvie (4:298) gives: "The stream of water which runs from the mill after it has been applied to produce the motion to the wheel." The Sc. term has the same definition.

tail-stream (Sc.) See tail-dam.

tail-water (E) See tail.

tain (Celt., OGaul & SGael.) Chalmers (26) gives the Celt. and OGaul. as: "river" and Dwelly (925a) the SGael: "water". The PN Tain in Scot., is given by a number of authorities under this term. See Ross (208).

tairiden (Ir.) This word is listed in eDIL – ALI (4:218.29) gives: "thairidin", and (220.15) "tairidne". The word is all to do with leading water for a specific purpose in some way, such as a watercourse, mill-race or ditch.

þeóte (OE) B-T (1053) give: "*A pipe or channel through which water rushes.*" WW (1:147.34 & 488.22) give "*canalis*" and "*tubo*" respectively. See wæter-þeóte and EPNE (2:203) for related PN's.

thorn-drain (E) An agricultural drainage term. Peacock (1889:561) gives: "Before drain tiles became common it was the custom among farmers to drain their land by digging trenches and burying sticks, commonly thorns, in them; these were called *thorn-drains*, and the process *thorndraining.*"

thorough (E) An agricultural drainage term. Stephens (5:249) states: "Some of the earliest written accounts of the treatment of the heavy clay-lands of these counties describe these ditches or "thorows" (hence has been derived our term "thorough-drainage") as being cut from 2 to 2 1/2 feet in depth. Some material was deposited which by its decay might leave a channel for the water to flow through." See water-thorough.

thorow (E) See thorough.

thour (Co.) See dour.

þroc (OE) See drock and EPNE (2:213-4).

through (E) An agricultural drainage term. From the OE *þrúh*, B-T (1073), "*a trough, pipe, conduit.*" Also in the form *ðrúh*, Lindsay (1921b:179 line 320). Lindsay (1921a:43) also gives another OE form "*thruuch*". WW (1:200.14) list the compound "*wæterþruh*" which corresponds with the L *caractis*. See throw and PNGl (1:120 & 4:181); Horovitz (592); Wallenburg (298); PNWe (2:295); PNWo (169) and PNYW (7:261) for PN connections.

throw (E) An agricultural drainage term. The Shropshire form of the word through, given by Jackson (440) as: "a hole cut through a hedge as a channel to let water run off the land."

thruff (E) See trough.

thrugh (E) See trough.

ðruh (OE) See through and wæterþruh.

þrúh (OE) See through.

thruuch (OE) See through.

þurruc (OE) See drock and thurruck.

thurruck (E) Apparently similar to pennock and pinnock. Wright (960b) gives: "A drain" and Parish & Shaw (173): "A wooden drain under a gate; a small passage or wooden tunnel through a bank." See EPNE (2:217).

tie (Co.) Jago (1882:292) gives: "Tie. A large wooden trough used for wasing ore", and (302) "Tye, the same as strek … but worked with a smaller stream of water." The Co. equivalent of the northern hush. See PNGl (4:180).

tile-drain (E) 'A drain made with tiles' – given by Stephens (5:263) and Ogilvie (4:374b).

tirle (E) Brook terms are suggested for tirle, trill and trull by Ekwall in ERN (323: 409-10 & 418). As tirle, trill and trull enter into at least six R or PN's, we should expect a common etymology, but, in this case, nothing is definite. See PND (634); PNO (469) and PNW (117, under Trowle Farm).

toe (Sc.) See tae.

tongue (E) Not normally defined as a watercourse but is shown as such at NGR SE0001.

top carrier (E) An irrigation term given in Wikipedia (under Water-meadow) and FD. See carrier.

topping-trench (E) See trench-royal.

torrens (L) WW (1:178.1 & 325.37), from two of the vocabularies, give: "broc" and "burna" respectively as the OE equivalents; E 'brook'. PP (50) list "torrens" under "brokwater".

torrentibus (L) WW (1:51.17) give: "streamum", which is listed in their earliest glossary of vocabularies, an *Anglo-Saxon Vocabulary of the 8th Century.*

tout (E) Brogden (212) gives "A "grip" or tunnel under a road." a grip is a watercourse so too then is a tout by this definition.

traen (Wel.) Pughe & Pryse (2:523a) give: "a drain, a channel for water." Other references are lacking.

trainnse (SGael.) Only in Dwelly (966b) as: "trench, drain."

tramites (L) WW (1:146.37) give: "wæterweg", from *Abbot Alfric's Vocabulary*; the term is still used in Italy, defined as 'conduit'.

traphont (Wel.) This term is listed in the GPC (3557b) defined as: "aqueduct", a later arrival – appearing in the 20th C. See dyfrbont, dyfrffordd and pont.

traversing-trench (E) See trench-royal.

travish (E) The EDD (6:228a) gives: "A little drain cut to carry off water."

traw (E) See trough.

treble-trench (E) See trench-royal.

trench (E) An agricultural drainage term from the OF *trenche*. Loudon (726:4412) gives: "The trench is a narrow shallow ditch, for conveying the water out of the mains to float the land." A trench in the IOM is shown on OSL 95 at SC3999.

trench drain (E) An agricultural drainage term given by Boswell (24) who states: "TRENCH DRAIN is always cut parallel to the trench, and as deep as the tail drain water will admit when necessary." See trench.

trench-royal (E) An agricultural drainage term – Duncumb (111) gives the following extract from the work of Robert Vaughan, who published an essay on the subject of watering in 1610: "... in running my maine trench, which I call my trench-royal. I call it so, because I have within the contents of my worke, counter-trenches, defending-trenches, topping-trenches, winter summer-trenches, double and treble-trenches, a traversing-trench with a point and an everlasting-trench". This extract contains an amazing list of eleven irrigation terms for watering meadows that are unlikely to be found anywhere else in print; the definitions of which, we may never know.

trend (E) According to Wright (1880:977a) a trend is a stream in Dev.

trenk (Sc. & Shet.) Warrack (628b) gives: "a narrow, open drain", and, (629b): "Trink, Trinck, a trench; a narrow, open drain for the passage of water." Angus (144) also lists trink as: "a narrow trench."

trent (E) The Trent, a watercourse which flows through Liverton in Yorks., is a long ditch of water, shown on OLM 26 at NZ7114. The OED1 for trend (324b) gives: "A rounded bend or circuit of a stream, dial." but nothing on trent other than to mention that trent is a variant of trend with meanings all to do with bending and circuitry. See trend.

triangular-drain (E) An agricultural drainage term. Ogilvie (2.96) lists: "triangular drain." Loudon (709, fig. 652) has: "the wedge or triangular sod drain." See wedge-drain.

triangular sod drain (E) An agricultural drainage term. An alternative name for a triangular drain given by Loudon (709).

tributary (E) A 'branch of a river, a side stream'. OSL 166 at TL3523 and OSX 193 at TL2831 both show one and OSX 194 shows five.

trickle (E) The OED1 (348a) gives: "a small fitful stream." See tricklet and trike.

tricklet (E) A very tiny stream, the dim. of trickle – Ruskin (244) gives: "a tricklet here at the bottom of a crag ...".

trig (E) Jackson (453) gives, for *trig* (2): "a small gutter, — same as Rigol (1)." Hartshorne (600) and Darlington (408-9) give the same and Lawson (1884b:16) gives: "trench." Also found compounded with gutter. See prick-gutter where trig is confirmed as an agricultural drainage term and PNWe (1:xiv 30) and (2:294).

trig-gutter (E) See prick-gutter and trig.

trike (ME) Morris (1865:84, line 2947) from *The Story of Genesis and Exodus* gives: "ineuerlic trike", explained in the (Notes, 161) as: "a rivulet, small stream, evidently connected with the verb trick-le."

trill (E) See tirle.

trinch (Sc.) DOST give this as: "A trench for draining off water, a drainage-ditch." It's a very similar word to trink.

trinck (Sc.) See trenk and trink.

trink (OrkNn.) Marwick (167a), gives: "narrow creek or channel of water, from Scots *trink*."

trinket (NIr. & Sc?) Patterson (108) gives: "Trinket, a small artificial water-course." Although this is obviously a dim. of the Sc. trink, Angus (144), one of few to record it, does not define it as a watercourse term.

troch (Sc. & Shet.) A variant of trow. Angus (144) gives the Shet.: "The shoot of a watermill."

trog (OE) B-T (1015) give: "*a water-pipe, conduit*" and associate it with mylen-trog, "trough". PNLa (18) gives: "O.E. *trog*, trough, later also a hollow or valley resembling a trough; bed or channel of a stream." See mill-trough and trough.

trone (E & Co?) Marshall (1873b:75) gives: "trench or drain", and Jago (1882:299): "a small furrow, or narrow trench". See droke.

trot (Co.) Pryce defines this term as: "the bed or channel of a river." The only other reference comes from Jago (1887:14b) which confirms Pryce's entry.

trouch (OSc.) A variant of trow.

trough (E & Sc.) From the OE *trog*. Kersey gives: "a pipe for the conveyance of water" and Cunliff (90) has: "thruff, thrugh, a walled drain for carrying away water." Crossing (1:50) mentions Long-a-Traw, on Dartmoor, which he defines as : "a long trough", which can be seen on OSOL 28 at SX6864. He also mentions (1:60) "Henchertraw, where the stream is shut in between high banks." For the Sc., Warrack (631b) gives: "Trough, the wooden conduit conveying water to a mill-wheel." See trog and the MED.

trovel (E) Wright (1880:980b) defines this as: "A mill-stream." Perhaps connected, in some way, with trov, "a trough" given by Peacock (1889:578).

trow (E & Sc.) A variant of trough; a term which leans more towards Sc. The SLB (1908:L.) gives trow as a watercourse term and in the SLB (1906:360:13) it equates with 'conduit'. The Durham Roll (3:979 & 985) has similar and Norwich (1906:46 & 321) mentions "Trowys Eye" and "Trous Ee", respectively, but here, as a name of a watercourse, as ee is the generic. For the Sc., Jamieson (4:629b-30a) gives: "The wooden spout by which water is carried to a mill-wheel ... it is also called a shot." Warrack (632a) gives: Trow, a trough; a conduit carrying water to a mill-wheel; the lower ground through which a river runs", and, "Trowse, Trows, the conduit carrying water to a mill." The DSL list the OSc. variant, trouch (DOST) and the Sc., troch (SND). See shot.

trows(e) (Sc.) Variants of trow.

trull (E) See tirle.

trunk (E) Ogilvie (4:449a) gives: "A trough to convey water from a race to a water-wheel." Cope (97) states: "an arched drain under a road" and Holland (262) under "PLAT", mentions that: "About Frodsham the watercourse itself under the *plat* is called the *Trunk*." See trunk-drain and trunk-way.

trunk-drain (E) An agricultural drainage term given by Rees (33); one which is: "laid across the bed of a river."

trunk-way (E) Rye (235) following Forby (2:358) gives: "A watercourse through an arch of masonry, turned over a ditch before a gate. The name arose, no doubt, from the trunks of trees used for the same purpose in ancient and simpler times, and even now, in the few wooded parts of both counties."

tturraeth (Ir.) See turraeth.

tubo (L) Lindsay (1921a:43) gives: "thruuch", and (1921b:179, line 320): "druh", both 'water-pipe, conduit'. See through and þrúh, B-T (1073).

tuiridne (Ir.) This appears in ALI (4:166.24). See tairiden with which it is associated (eDIL entry) and turraeth.

tunnel (E) Gwilt (1048) gives: "A subterranean channel for carrying a stream of water under a road." Carr (2:221) gives:

"An arched drain."

tunnel-drain (E) Marryat (329-30) mentions this term twice, apparently used of a watercourse from houses used as manufactories.

turf-drain (E) One of many agricultural drainage terms given by Ogilvie (2:96a). Dickson (1805-7:1:332) refers to this term as a "shouldered turf-drain."

turloch (Ir. & SGael.) This is the equivalent of the OE winterburnan, E winterbourne – a brook, wet in winter and dry in summer. The Irish dictionaries of Coneys, O'Brien and O'Reilly all agree as does Dwelly (984a) for the SGael.

turraeth (Ir.) O'Donovan (1856:1:262 line 12) gives: "tturraeth", which translates 'mill-race'. See eDIL.

tye (Co.) See tie.

U

uaeterthrouch (OE) See wæter-þrúh.

uarach (SGael.) Not in Dwelly as a watercourse but he gives (988b) uar, 'waterfall' and uaran, 'fresh water'. Two examples are shown on maps, one on OSX 460 at NB4252 and another on OSX 460 at NB4949. Joyce (1869:1:453) mentions uaran and gives a number of relative PN's in Vol. 3. Hogan (693a) mentions Uar as the name of a stream, which may be a contraction of uaran.

uaran (SGael.) See uarach.

uena (L) WW (1:307.13) give: "*uena*, aeddre"; 'vein'. See ǽdre.

ùidh (SGael.) Dwelly (991b) gives: "slow running water between two lochs." CR (10:281) mentions that uidh: "is better spelled *aoidh*," For mapped examples of uidh, see OSX 434 at NG8985, OSX 441 at NH6396 and OSX 445 for 6 more examples of ùidh running between lochs.

uillt (SGael.) A variant of allt – Dwelly (26a) gives: "mountain stream, rill, brook." A number of instances have been mapped; one on OSX 355 at NR4866 and another on OSX 433 at NG8475, further, there is a nuilt shown on OSX 363 at

NS1368 and many references and other forms of this word can be found in CPNS, all listed in Index (E). See allt.

uinterburna (OE) See winterbourne.

uisc (OWel. See wysg.

uisg(e) (Ir. & SGael.) For the Irish, O'Reilly (550a) gives: uisg and uisge, "water, a river." For the SGael., Dwelly (995b) gives: "uisge, water, river, stream." A 'uisg' is shown on OSX 353 at NR2863; a 'uisge' on OSX 429 at NH0038 and even a 'uisage' on OSX 373 at NM5324. Many references and other forms of this word can be found in CPNS, all listed in Index (E).

uisgeachan (SGael.) Dwelly 995b gives: "river, stream." The contracted form, uisgeacha, is shown on OSX 375 at NM5830.

uisge-ruithe (SGael.) See ruith.

uisgrian (Ir & SGael.) O'Reilly (550a) gives: "an aqueduct." Dwelly (995b), for the SGael., the same.

uisin (Ir.) The primary sense of this term is 'water' but a stream definition also exists and is given as such by eDIL.

under (E) Brogden (216) states: "Under, under-grip. – A drain or "sough" underground", and The EDD(6:306b) gives: "-grip or -grup, an under-drain; a concealed watercourse in wet soils." See the under- compounds.

under-ditch (E.) An agricultural drainage term given by Vancouver (1795:203). See land-ditch.

under-drain (E) An agricultural drainage term given by Dickson (151) and Peacock (1889:535, under suff). Forby (2:363) defines under-grup as an "under-drain".

under-grip (E) See under.

under-grup (E) Rye (238) gives: "An under-drain, a concealed watercourse in wet soils." See under.

upe (Lett. & Lith.) See avon.

upright drain (E) Rees (33) gives: "This upper drain will cut off the water, when it rises to the surface; while the upright drain will convey it to that cut along the lower side of the wet ground, where the water will subside into the porous subsoil."

ur(a) (Bas.) See owr.

uuaeterthrúch (OE) See wæter-þrúh.

uuaeterđruh (OE) See wæter-þrúh.
uuaterþruh (OE) See wæter-þrúh.
uueterþrúh (OE) See wæter-þrúh.
uy (Co.) See guy.

V

vadel (SNn.) Jacobsen (1928:1017b) gives: "occasionally, in sense of channel in the water from a land-locked bay flows into the sea."
vaedik (Shet.) Angus (150) gives: "a channel; a small stream or the bed of a small stream."
vedek (SNn.) Jacobsen (1928:1037a) gives: "a ditch for draining off water; not common in this sense", and also the variants "vetek, vettek, videk and vjedi."
vein (E) A channel or run of water for the purpose of washing minerals in a mine, much the same as a hush. There is a disused vein near to a quarry shown on OSOL 31 at NY8542.
vet(t)ek (SNn.) both variants of vedek.
videk (SNn.) A variant of vedek.
vjedek (SNn.) Jacobsen (1928:1059b) gives: "a ditch (for draining off water), brook, etc ...."
vjedi (SNn.) A variant of vedek.
vliet (Dut.) See fleet.
vlot-gutter (E) See flot-gutter.
vy (Co.) See guy.

W

waefleed (Sc.) Warrack (649b) gives: "the water of a mill-stream after passing the mill, the away-flood." See wamflet.
wælla (OE) See well.
*waesse (OE) See wash.
wæter-ǽdre (OE) B-T (1160) give: "*A vein of water, a spring.*" See ǽdre.

wæter-burne (OE) B-T (1161) give: "*A stream of water.*"

wæter-furh (OE) B-T (1162) give: "*A trench.*"

wæter-ganc (Dut.) See watergang.

wæter-geláda (OE) WW (1:339.4) give: "*wætergelada,* aqueductum." B-T (1161) give: "*A water-way, an aqueduct.*"

wæter-gelǽt (OE) WW (1:211.13) give: "*wætergelæt,* colimbus" and B-T (1161) give: "*A water-course, an aqueduct.*"

wæter-gyte (OE) B-T (1161) give: "*A pouring of water, a water-course.*" See gyte.

waeteringe (Dut.) See watering.

wæter-leyd (Dut.) See water-lead.

waeter-loop (Sc.) See loop.

wæter-riþe (OE) B-T (1161) give: "*A stream of water.*" See riþ(e) and rithe.

wæter-scipe (OE) B-T (1161) explain it: "*a body of water, a piece of water, water.*" WW (1.184 line 12) give: "*colimbus, aquaeductus, wæterscipeshus*", further, in (1.211 line 13) *colimbus* is equated with *wætergelæt,* which B-T give as: "*a water-course, an aqueduct.*" The findings here suggest that *wæterscipe* is definitely an OE generic, confirmed by the entries in EPNE (2:238), PN Ess (592), PN Hrt (127 & 304) and PNW (394). See water-sipe.

wæter-slæd (OE) B-T (1162) give: "*A valley with water in it.*" See PNBrk (3:737) and slade.

wæter-streám (OE) See water-stream..

wæter-þeóte (OE) WW (1:118.11), under aquagium, give: "*wæterþeote*". B-T (1162) give: "wæter-þrúh", for this term. See aquaeductus and wæter-þrúh.

wæter-þrúh (OE) B-T (1162a) give: "wæter-þrúh, *a water-pipe, conduit ...*" and the forms: "uueterþrúh", "uua[e]terthrúch" and "uaeterthrouch". Lindsay (1921b:32) gives: "*uuaterþruh*"; and (1921a:110), "*uuaeterthruch*"; He also gives *uuaeterðruh* (1921b:289b). Interesting forms which use double *u,* before it became *w.* The term does not appear to have come down to us in ModE. See through, throw and wæter-þeóte.

waeterðru(u)m (OE) See canalibus.

wætertige (E) See wæter-tyge.

wæter-tyge (OE) B-T (1162) give: "*An aqueduct* :-- Wætertige, *aquaeductus, canalis.*" See PNGl (4:183).

wæter weg(e) (OE) See waterway.

wa-gaen (Sc.) Warrack (650a) gives: "the channel in which water runs from a mill."

wa-gang (Sc.) Warrack (660b) gives: "the channel of water running from a mill. Cf. Wa-gaen."

wait (Sc.) From ON *veit*, 'a ditch'. Warrack (651a) gives: "the water-course from a mill." The EDD (6:362a) state that it is: "Also written wate." The DSL (SND) gives the same.

waiter-foor (E) See water-foor.

waiter geeat (E) See water-gate.

waiter-tyeble (E) See water-table.

wall (E) Normally defined as an embankment to hold back river or sea water but also a watercourse, shown as such at NGR TG4219.

wall-drain (E) An agricultural drainage term. A 'main-drain'. Arkell (325) gives this term in his paper *On the Drainage of Land*, stating that: "Stone drains are various; the most common here are wall and dribble, or rubble, the former as main, the latter as tributary." Loudon has a sketch of 'walled drains' on page 707 (fig. 643a & c), naming them as: "the walled or box drain" and "the walled or the triangular drain" respectively. Ogilvie (2.96a) also lists: "walled or box drains." See box and triangular drain.

walle stream (ME) Layamon (line 2849) gives this compound; the ME form of the OE *wille-stream*. See well- and wille-.

wall-spring (E) An agricultural drainage term. Dickson (1805-7:1:290) is one of the few to mention it: "... so as to intercept the wall-springs and land-floods ...." Marshall (1796a:388) gives: "WALLSPRING ; a cold, wet, springy, or spewy part of land."

wamflet (Sc.) Warrack (653b) gives: "the water of a mill-stream after passing the mill. See wamfleed.

ware (E) See weir.

warping-drain (E) A warping drain brings silt from a tidal river to gradually fill and reclaim land. One is shown at NGR SK772975.

warstead (E) Only in Williamson (104) as: "a watercourse".

wash (E) Clarke (304) gives: "When those fens were first embanked and drained, narrow tracts, called "dales," or washes, were left open to the river...." A wash is shown on OSX 177 at TL8703 and washes on OSX 235 at TF1619. See ERN (436-8) for RN's and the connection with OE *waesse*. See wysg and the MED.

washaway (Co.) See washway.

wash-mill (E) Lisle (317) states: "to wash our sheep on the morrow, I asked my shepherd, what time in the morning he would drive them to the wash-mills." Other refernces are lacking. See wesh-beck.

washway (E) Possibly, 'a fordable stream', one flowing across a road or track. AR (1793:66) states: "the North mail cart, going through Tottenham Washway, was under water ...." Possibly connected with washway is Washaway and Washaway Farm in Cornwall, both of which can be seen at NGR SX0369, although here, there is no indication of a stream crossing the road. Bannister (1871:187a) gives: "entrenchment (*fos*) near the *way* or road." See Padel (177); Watts (653b); Weatherhill (82b) and for washway in field names, PNO (1:182; 2:332 & 470).

waste drain (E) Hatfield Waste Drain is shown at NGR SE7510, it flows into Three Rivers, which joins the R Trent. It connects with other drains in the area and not used for foul water discharge.

wate (Sc.) See wait.

water (E, Ir., ME; Nir., O & ModSc.) From the OE *wæter*, B-T (1160). '*A stream or river*', very often used as a prefix or suffix for RN's, common in Scot., less so in Eng. and also found in Irl. Dinsdale (143) gives: WATTER, Water; also a river." Elworthy (821) gives: "water, a stream; brook. (Very com.)." Very common in Som. that is. A mapped example is shown on OSX 173 at TQ1799. The ME form watere comes from Layamon (1:86). In Irl., OSNIDS 12 shows Fairy Water at H2578; The Black Water at H2873 and water on OSNIDS 4 at C8125, for the north and OSIDS 50 at N9339 and O1049 shows southern examples. As already mentioned this term is very common in Scot., Warrack (657b) gives: "Water, Watter, a river, a good -sized stream …", and there are many more references to be found in The DSL (SND). Water of Ae is shown at NGR NY058849; another water at NR6206 and there is even one on the Shetland mainland, shown on OSX 466 at HU4019. The OSc. form is wattir given by DSL (DOST) an entry well worth perusing. See EPNE (2:238, under wæter); the MED and The OED.

water beck (Sc.) This is a rare tautological term, shown at NGR NY2480 and, as Johnston (245) points out, "water and beck both mean brook." See Ross (2001:221).

water bridge (E) 'An aqueduct or a bridge which carries a watercourse over a canal', such as Tunstall (Dunstall) Water Bridge, which carries a brook, here, the Smestow Brook, over the Staffs. & Worcs. Canal near Wolverhampton.

water brook (E) Robinson (1855:23) gives this term as a definition of burn.

water-carriage (E) An agricultural drainage term given by Cope (14) under carriage and mentioned by Davis (118) who gives: "This produced by degrees that regular disposition of water-carriages and water-drains which, in a well laid out meadow, bring on and carry off the water …." See carriage.

water-carrier (E) Dartnell (23) gives: "Carrier, Water-carrier. A large watercourse." Atkinson (1891:410) mentions water-carrier too.

water-cast (Sc.) The DSL (SND) gives: "a water channel, ditch." See cast.

water channel (E) Cornish (129) gives: " After such fogs, though rain may not have fallen for a month, and there is no water channel or spring near the dew pond, the water in it rises prodigiously." Stephens (1889:1:75) mentions: "water channels in ploughed land."

water condite (ME) See water conduit.

water conduit (E) Polychronicon (4:365) gives the ME form: "… and i-made a water condite in to his owne hous …"; the variant, water conduyt, appears in other Mss. Godstow (45-6) gives: "water cundit"; which would appear to have been used more in the sense of water supply rather than as an independent watercourse. See, especially p 46, for an in depth discussion on the whole operation and the MED for other references and a full list of forms.

water conduyt (ME) See water conduit.

watercorse (ME) See watercourse.

watercourse (E) A watercourse is defined in the Land Drainage Act 1991 as: *"watercourse includes all rivers and streams and all ditches drains, cuts, culverts, dikes, sewers other than public sewers within the meaning of the Water Industry Act (1991) and passages , through which water flows."* The Act does not state that water must flow through the watercourse at all times to be a watercourse. Glasscock (31) gives some ME variants: "watercorse"; (140) "wattercours" and (141) "wattercourse". His are the earliest known references for the watercourse term and the SLB (1906) also gives a number of ME variants. A mapped example can be seen on OSX 286 at SD4324. See the OED.

water cundid (ME) Wycliffe (2:220a, 3:Kings:32) gives: "and he made a water cundid ...." See water conduit.

water cundit (ME) See water conduit.

water cundiyt (ME) Wycliffe (2:94b, 2:Kings:24) gives: "to the litil hil of thed water cundiyt ...." See water conduit.

watercut (E) Cornish (9) gives: "Down every furrow, drain, watercourse, ditch, runnel, and watercut, the turbid waters were hurrying ...." This term is shown on OSOL 21 at SD9437.

water-ditch (E) An agricultural drainage term which can be found in Marshall (1787:277) "The grazing parts are divided into inclosures, of various sizes and figures, by means of water-ditches, of different widths, from five or six to eight or ten feet wide." See ditch.

water-drain (E) An agricultural drainage term. Davis (118) gives: "This produced by degrees that regular disposition of water-carriages and water-drains which, in a well laid out meadow, bring on and carry off the water ...." Ogilvie (602c) gives: "A drain or channel for water to run off." See drain.

watere (ME) See water.

waterfall (E) Part of a watercourse which falls rapidly from a height; sometimes a great height, such as Angel Falls in Venezuela, which is the world's highest at 979 metres – 3212 feet – almost as high as Snowdon! The highest waterfall in GB is Eas à Chual Aluinn, shown at NGR NC2827, in Scot. at 200 metres – 658 feet.

water-fence (E) An agricultural drainage term which can be found in Marshall (1787:278) "These water-fences, running in all directions, and being of various widths, makes it probable that the principal part of them were the smaller furrows, or partial drains, which carried off the rains, back-water, &c. in a state of nature." See fence ditch and –drain.

water-flod (ME) See flood.

water-foor (E) An agricultural drainage term. PP (517) gives: "water fore". Heslop (771) gives: "Water-foor, Waiter-foor, a furrow made by the plough to drain off surface water." See water-furrow.

water fore (ME) See water-foor and water-furrow.

waterforowe (ME) See water-furrow.

water fur(r) (Sc.) An agricultural drainage term. The Sc. form of the E water furrow. Warrack (658a) gives: "a furrow made to drain off surface-water." Gregor (207) defines it under the Banff compound, "Wattir-fur, a furrow in land to draw off the water." Jamieson (4:745b) lists it too. See The DSL (SND) under water.

water furlong (E) AD has a number of ME forms: "Waterfurlange or Le Syke" in Lincs. (1.347), "Waterfurlunge" in Warks. (2.241) and "Waterifurlung" in Northants. (2.471). There is, in Stamford, Lincs., a road named Water Furlong which is actually well over a furlong (about 200 metres) in length. The modern map does not show any form of channel connecting what may have once been an ancient 'water furlong' to the nearby river, but, the frequency of this term in the various counties, plus the fact that a "Waterfurlange or Le

Syke" existed towards the end of the medieval period, suggests that 'water furlong' was used as a fairly common agricultural draining term, during the ME period, much the same as water furrow. See wet furlong.

water-furrow (E) This term is listed in the PP (517) as: "water fore", L, "sulcus", and (734) "waterforowe"; the AD (1:335) as: "weteforrewis" and the CLB (510) as: "watir-fforough". However, the earliest mention comes from an A-S charter (BCS 477) dated 854. Peacock (1867:91) gives: "Watter-foore (Pr. of water furrow), n. a gutter or open drain, often made with the plough", and Cope (99) gives us the Hampshire rendering: "water-vore". See water-foor; PNNth (273) and, for the Sco. form, waterfur(r).

water-furrow drain (E) An agricultural drainage term given by Rees (38).

watergagium (L) Martin (342b) gives: "a watercourse" and "waterganga: watercourse; an aqueduct" plus watergangius: "trench to carry off water; an aqueduct." See watergang.

water-gait (E) See water-gate.

watergang (E, Fl., OSc. & Sc.) All gang compounds are from the ME *water* plus *gang*. Kersey's *Dictionarium Anglo-Britannicum*, published in 1708, is one of the earliest to define this term as: "a trench or course to convey a stream of water." Peacock (1889: 601), quoting from the *Dictionarium Rusticum*, 1726, gives: "*Water-gang*, a trench, trough, or course to convey a stream of water, such as are usually made in sea-walls to discharge and drain water out of the marshes" and Macray (145) gives: "Wetergangis and watergangis, water-ways, drains". Battle (22) adds another instance: "a water-gang, of sixteen feet." For the Sc., Jamieson (4:745b) gives: "The race of a mill" and Warrack (658a) "a mill-race; a water-course; a right to draw water along a neighbour's ground to water one's own." The OSc. form is "wattirgang(is)" and "wattir-gang", given in the *Charters of the Royal Burgh of* Ayr (121) and the *Registrum Magni Sigilli Regum Scotorum* (2:74) respectively; the latter also gives: "wattergang" (2:91). There is also a Dut. connection – Kiliani (648b) gives:

"wæter-ganc, Aquagiu*m*, aquae ductus" and USBG (Bel.:ivb) list the term as "drainage ditch, stream". See watergagium; The DSL (DOST); the MED; PNYE (323).

waterganga (L) See watergagium.

watergangis (E) See watergang.

water-gate (E, OSc. & Sc.) Heslop (771) gives: "a dry stream bed." Nodal (278) uses the form "water-gait, a gully or reft in the rock, which in summer is the bed of a streamlet, but in winter is filled by a torrent." The EDD (6:Supp., 177) states that it is written "waiter geeat" in West., and OSOL 41 at SD8242 shows one. Jamieson (4:745b) does not define watergate as a watercourse, but there are plenty of references in the The DSL (DOST), under Wattir gate and The DSL (SND), under water. See the MED and OED1 (10:170c).

water-gully (E) Addy (277) gives: "watter-gully, a watercourse."

water gutter (E) Robinson (1855:45, under dike) gives "the bank supporting the hedge or fence along the bottom of which there is a runnel or water gutter." See PNRu (194).

watering (E) Modern maps show PN's that suggest the possibility of what were once waterings, usually where a brook crosses a road at a low level beside a ford, such as those shown at NGR TM1492 and TM3859. In Kent there is a Five Watering Sewer at TQ9424. The waterings, other than the Kent one, suggest that they were watering places rather than a term for the watercourse there flowing, but, a watercourse definition is possible too – Jennings (1907:4) states: "It is necessary to keep much to the right, and pass an old farm before striking off towards the grey and frowning old mass of masonry, for the "waterings" are wide and numerous, and it is very easy to lose a good deal of time and trouble on these marshes", waterings here are some sort of watercourse for draining a marsh, a term confirmed in The EDD(6:403b). Chaucer (103 line 826) states: "Unto the waterynge of seint Thomas", explained on the same page as "a brook at the second milestone in Old Kent-road." This, of course, may mean that the brook carried the name St.Thomas' watering or was just a place on the brook for refreshment. The Dutch etymologist, Cornelii Kiliani, in his *Etymologicvm Tevtonicæ Linguæ*, (650), published in 1599, gives: "waeteringe – aqua, amnis, fluuius, flumen", which has come down to us since his time as 'wetering', a term fairly common in the Nth. today for a watercourse or draining ditch, which is also listed in USBG (Nth. fwd:b). Sweet Waterings, a watercourse in Gl., is shown on old maps and can be seen in the relative square at NGR ST6124988861 on Geograph. See PNGl (3:115).

water kundit (ME) Wycliffe (2:94a, 2:Kings:24) gives: "and thei camen to the hil of the water kundit ...." See water conduit.

waterladde (ME). See water-lade.

water-lade (E, ME & OSc.) The earliest reference for this term would appear to be given by WW (1:339.4), in their 11th C *Glosses, Latin and Anglo-Saxon* as "*wætergelada* – aqueductum." Other references are given in the OED1 (10:174a). AD (2:427) has the ME form "Watersclade" and the CLB (28 & 31) gives: "waturlad" and "waturlade" respectively. Although very few, OSc. references are given in The DSL (DOST), under Wat(t)ir, and a mapped example is shown at NGR TF4561. See the MED (under water-lode) and water-lead, lade and lead.

water-lead (E?, Ir. & Sc.) Chiefly an Ir. and Sc. term, the OED1 (10:174b) does not give any E definitions. There is also a Dut. connection – Kiliani (279b & 648b) gives: "water leyde" and "wæter-leyd" respectively, Aquae ductus, aquagium." See water-lade, lade and lead.

waterleat water leat (Co. & E) See leat.

waterlete (E) AD 3:560 (D.1287), a Somerset deed, is the source for this rarity: " la waterlete." Kiliani (648b) gives: "waeter-laet. Aquaeductus, riuus"; the cognate Dut. form.

water leyde (Dut.) See lead and water-lead.

water-lode (E) This term is to be found in the catalogue of Battle Abbey documents dating back to 1512-15 (Battle 134-5): "Scota, or Assessments, made at the Sessions of the Waterlode, in Hoo-Marsh, April 26, 1512, for the Scouring and Repairing the said Waterlode"; mentioned again in 1515.

waterloop (Sc.) See loop.

water-passage (E & OSc.) Grose (73) uses the term water-passage in his definition of gote, as does Phillips, under wydraught. It is used in OSc. and is equated with water-gang.

water-prill (E) Duncumb (109) mentions the work of Robert Vaughan, published in 1610, and gives the following extract: "... containing the manner of summer and winter drowning of medow and pasture by the advantage of the least river, brooke, fount, or water prill adjacent ...." This is, perhaps, the only mention in print.

water-race (E) The AR (90b) mentions this term, which is shown on OSOL 26 at SE5386 and OSOL 31 at NY9919. See race.

waterrene (E) This is a rare term and can be found in the AD (2:350), in a deed dated 1374.

waterrflod (OE) See flood.

water rin (Sc.) The DSL (SND) give this and –run defined as: "a runnel of water, a surface drain or gutter for carrying off water, a streamlet."

water run (Sc.) See water rin.

water-run (E) This agricultural drainage term is listed by Stephens (1889:1.121) as an alternative for gaw and gives: "Gaws or water-runs should never be neglected to be cut after lea ploughing," See gaw.

water-runner (E) Cornhill (62:387) gives: "Take the first runner you come upon for guide, for a watter-runner will always lead you to the bottom."

water-scape (E?) Rees (Vol. 38) gives: "WATER-*Scape*", and says that this word is: "the Saxon *waterschap*", denoting "an aqueduct, drain or passage for water." However, waterscape or waterschap is not to be found in B-T, the nearest is wæterscipe (1161) explained as: "*a body of water, a piece of water, water.*" WW (1.184:12) give: "*colimbus, aquaeductus, wæterscipeshus*" and again no entry for waterscape or waterschap is to be found, further, in (1.211 line 13) *colimbus* is equated with *wætergelæt*, which B-T give as: "*a water-course, an aqueduct.*" The findings here suggest that *waterscipe* is an OE generic, but water-scape is doubtful, however, the Rees *water-scape* could have been taken from a source such as Kiliaan who gives, (649b): "*waeter-schap*, aquagium, aquæductus." Köbler (OSax.) gives: "*watriscapum*" (an OSax-Latin hybrid) as NHG "wasserlauf, NE watercourse", which would make water-scape a true generic. Waterschap is a term used in the Netherlands for a district water board who manage a body of water. The word in the Netherlands has obviously come down to us with a different meaning to that given by Kiliani, but it's interesting to note that it is still used of a water feature. In view of the foregoing, it is obvious that water-scape should be classed as a qualifier.

water schedellys (ME) See shadel; shedele and the MED under water-sheteles.

watersclade (ME) See water-lade.

water-sheteles (ME) See waterschedellys.

water-shoot (E) Mortimer (2:207) gives: " ... and the water-shoots of other adjacent lands."

water-shut? (E) Hazlitt (1:119) gives: "... or for some water-shut."

water sike (E) A watercourse running from Bunton Hush, shown at NGR NY9401. Hush is a lead mining term, used for the purpose of washing minerals, common in the vicinity of Bunton and elsewhere in the north. See hush.

water-sipe (E) Robinson (1876a) gives: "water-sipe, the course in which the water soaks through the ground to supply a pond or well." See sipe.

water-stead (E) Halliwell (2:119a) gives: " The bed of a river", discussed by Atkinson (1868:495). See water-stell.

water-stell (E) Atkinson (1868:495) States: "Stell, a large open drain in a marsh ... There can be no doubt that this is merely the abbreviation of water-stell (water-stead, the bed of a river, Hall, still exists)." See stell.

water-still (E) Heslop (772) gives: "a channel for water flowing through a marsh." See water-stell.

water-stream (E) From the OE *wæter-streám*, B-T (1162).

Spenser (293b:XX1) gives: "*Like as a water-streame, whose swelling sourse shall drive a mill, within strong banck is pent. And long restrayned of his ready course; so soone as passage is unto him lent ....*" The *Ormulum* (line 18092) gives: "*waterrstræm*" and Smiles (3:20) gives the ModE form.

water-table (E & Sc.) Elworthy (821) gives: "The ditch on each side of a road; also a small hollow made across a road to carry off surplus water." Heslop (772) gives the dialectal: "Waiter-tyeble, a water table." For the Sc., Warrack (658b) gives: "the ditch or gutter on each side of a road for carrying off water." It is a term mentioned in many other word books always with the general meaning: 'a ditch at the side of a road'. See tabling.

water-thorough (E) Ellis (23) and Britten (41) both give this agricultural drainage term. See PNNth (273) and thorough.

water trench (Sc.) A rare term, Inchaffray (175) gives: "water-trench (stagnum)."

water trough (E) The OED1 (10:157c) list this term, dated to 1667. Wycliffe (1:123a, from Genesis (24:20) gives: "water trowis" and, (142a): "water trowes" from (30:38) and, "watyr trowis" from (30:40). The Durham Roll (3:985) also mentions watertrowe in its Glossary.

water trowes (ME) See water trough.

water trowis (ME) See water trough.

water-trunk (E) An irrigation term – given by MR (234): "I have found it of great use to bestow a watering on my fields, by means of water-trunks."

water-vore (E) See water-furrow.

waterway (E & NIr.) From the OE *wæterweg*, given by WW (1:146.37). B-T (741 & 1162) define it: "*A water-way, a channel connecting two pieces of water.*" BCS 963 (S622) gives: "*wæter weg*" and "*wæter wege*" – the latter is repeated in KCD (1198). PP (518) equates water wey with the L *meatus*, 'course' and waterwey is mentioned in three Surrey Grants: (AD 2:454, 484 & 585) and Glasscock (16 & 42) gives the same form. Battle (178) gives: "*water waye*" and a mapped example is shown on OSX 285 at SD3717. A NIr. one is shown on OSNIDS 27 at

H2920. See PNBrk (3:666) and the MED under water-wei.

water waye (ME) See waterway.

water-wei (ME) See waterway.

waterwey (ME) See waterway.

wath (E?) A wath is normally defined as 'a fordable stream', however, the wath shown on OSOL 5, at NY3237, is applied to the whole stream as if it were a generic term and not just to a particular crossing point on it.

watir-fforough (ME) See water-furrow.

watter (E & Sc.) See water.

wattercours (ME) See watercourse.

wattercourse (ME) See watercourse.

watter drawcht (Sc.) The DSL (SND) give this compound from a 16th C reference, the Sc. form of the E simplex – draught.

watter dyke (E) Watter for water is fairly common in some northern dialects. The fact that watter is compounded with dyke here suggests a ditch filled with water, but, not necessarily running water. Dickinson (356) gives: "a ditch or sowe wide and deep enough to form a fence." Robinson (1855:189) has a completely different definition: "the worn holes in the roads or streets filled by the rain." Dickinson's 'sowe' (306), which is a Cumb. variant of sough, is described as: "a wide and watery ditch" – again, no indication of running water. Prevost adds (165a), in his *Supplement to the Glossary of the Dialect of Cumberland*, that: "sowe, 306. Is of artificial origin, generally for drainage purposes, in contradistinction to 'pow,' which is natural." It may have been used in the same sense as the Wirral farmers used fender, as a 'defender, protector' of lands, see Harrison (1898:85-6). Further evidence is lacking preventing a more thorough investigation into the real meaning.

wattergang (Sc.) See watergang.

watter gwoat (E) Dickinson (356a) gives one definition as: "A water gap in a fence."

watter strype (OSc.) The RGSS (6:418b) give this rare compound. See strip(e).

wattery lonnin (E) Dickinson (356b) states that a wattery lonnin is: "A neglected lane along which water is allowed to run, therefore, one which becomes a natural watercourse."

wattir (OSc.) See water.

wattir-fur (Sc.) See water fur(r).

wattirgang (OSc.) See watergang.

wattir passage (Sc.) See water-passage.

waturlad(e) (ME) See water-lade.

watyr trowis (ME) See water trough.

wayflame (Sc.) The DSL (DOST), under way, give this and waylaid defined the same: "a water-course constructed to serve a mill."

way-flude (Sc.) The DSL (SND) gives: "The outflow of water from a mill-wheel, the tail-race."

way-gang (Sc.) Warrack (660b) gives: "the channel of water running from a mill."

waygate (Sc.) Only in The (SND) as: "a means of drainage for surplus water."

way-goe (Sc.) Jamieson (4:752a) gives: "… place where a body of water breaks out". See gue and mill-gue.

waylaid (Sc.) See wayflame.

wayloom (Sc.) The DSL (SND) gives: "the tail race of a mill"; from a document dating back to 1734.

wear (E) See weir.

wedge and shoulder drain (E) Stephens (1889:5:262) gives a sketch of this agricultural drainage term and states that: "These were used in many parts of England where the soil was found to be firm enough to give durable turfs."

wedge-drain (E) An agricultural drainage term. Obviously the best type of drain for clay soils according to Burke (1841:29) who states: "and there is one, called "wedge" or "plug-draining," which has been within these few years rather extensively used in some of the strong clays; to which alone it is applicable, and better adapted to pasture than to arable land." Loudon (709, fig. 652) gives: "the wedge or triangular sod drain." See triangular drain.

weel (Fle.) See wheel.

weeth (Co.) This rarity comes from Bond who gives: "Stream – ..., weeth." No other authority appears to mention the term.

wefflin (Sc.) Warrack (664b) gives: "Wefflin, Wefflum, the water-course at the back of a mill-wheel."

wefflum (Sc.) See wefflin.

weir (E) From the OE *wer*. Bailey gives: "wear, ware ... or conveying the stream to a mill." There are a number of mapped instances – The Weir, near to Brockenhurst, Hants., on OSOL 22 at SU2801, is shown as a wet ditch and, in Warminster, Wilts., The Were, is shown on OSX 143 at ST8645. The Were on OSX 143 at ST8744 prefixes the Swan River and The Weirs in Burwell, shown at NGR TL5867, would appear to be a major collecting channel for all the local drains. Photographs of most can be seen on Geograph.

weir stream? (E) A term used by J. K. Jerome (135) in his book *Three Men in a Boat*, 'a stream that leads to a weir'?

welham (E) See whelm.

well (E) From the OE *wille*, and the later *wælla* and *wælle*. B-T (1228) give: "wille, wielle, welle, wylle, *a well, spring, stream, fountain*", and the compounds "wille-streám" and "wille-burne." Surprisingly, WW (1:178.9) give the modern form as early as the 10th C in the *Supplement to Alfric's Vocabulary*. Two wells are shown on OSOL 43 at NY8250 and three on OSOL 43 at NY6851. Three pages are devoted to this very widespread element in the EPNE (2:250-3), which has full lists of regional variants and associated PN's, making it well worth perusing. See the MED under wel(le).

well-drain (E) An agricultural drainage term. Ogilvie (4:618a) gives: "a drain leading to a well." Rees (38) gives: "*Well-drain*, in Agriculture, that sort of vent or discharge for the wetness of land, which is constructed in somewhat the well or pit manner."

welle-spring (ME) A stream from a well, much the same as a well-stream. See the MED.

well-strand (Sc.) Jamieson (4:767b) gives: "A stream from a spring." and adds a quotation from *An Agricultural Survey of*

Peebles by Charles Findlater: "The designation of the smallest rill of water is a syke or a well-strand, if from a spring-well if the water is of quantity sufficient to drive a small waterwheel for light machinery, it is called a burn." Warrack (665b) gives: "Well-strand, n. a stream flowing from a well" and The DSL (DOST) give: "welstrynde."

well-stream (E) OED 1 (9:304) lists this term, from the OE *wylle-stream*. See the MED, under well-strem.

well-stripe (Sc.) Warrack (665b) gives: "Well-stripe, a well-strand."

well-strynde (Sc.) See well-strand.

wellum (E) See whelm.

wesh-beck (E) A term from Whitby in Yorks., given by Robinson (1876a:217b) who defines it as: "the brook where the sheep are washed." Another entry (168a), but not watercourse connected, gives "Sheep-wesh, a roofless enclosure of loose stones near a stream, in which sheep are gathered for washing and shearing." See wash-mill.

weteforrewis (ME) See water-furrow.

wetergangis (E) See watergang.

wetering (Dut.) See watering.

wet furlong (E) An agricultural drainage term similar to water furlong, wet furrow or water furrow; there is very little to chose between them. AD (2:524 & 5:230) has the form "wetforlong" and (4.101) gives "weteforlong."

wet furrow (E) An agricultural drainage term, the same as water-furrow but less well documented. AD (1:335) gives a Leics. deed which includes the term "weteforrewis" and a farm near Skelton in Yorks., at NGR NZ6718, still retains the name 'wet furrow'. PNNt (282) under furh, discusses wet furrow alongside water furrow.

wever? (E) This is given by Wright (1880:215a), Halliwell (924b) and Wilbraham (89) as a generic term for river. Leigh (224) gives "wever, weever", and states: "from the Wel. wy or wye, a river", his 'weever' comes from Drayton (2.67): *"Weever, the great devotion sings, Of the religious Saxon Kings, Those riverets doth* *together call, That into him, and Mersey fall."* Holland (385), with reference to Halliwell's definition, states: "I think there must be a misconception here; *Weaver* is the name of a particular river which flows into the Mersey at Frodsham, and, as far as I know, never means a river in general." There is another river Weaver in Dev. and a river named Wevery, which runs into the Wye in Brecknockshire. Rivers with a common stem often point to a generic etymology, but, in this case, there is no evidence to support this view, therefore, 'wever' must be considered doubtful as a qualifier. See PNCh (1.38); PND (16) for the RN's, and, for a lost name, PNW (4 under By Brook).

wham (E & Sc.) This is a dialect word which comes from ON *hvammr*, sometimes hwamm, 'a small valley', found mainly in northern counties associated with PN's or FN's, as well as Scotland. Prevost (193a) gives: "a marshy hollow, gen. with water." Warrack (667a) defines it: "flat glen through which a brook runs." Sc. entries tend to support a watercourse definition more so than E and the word has obviously been transferred to the stream names at some time in the past, surviving as such on OSOL 10 at SE0867 and OSOL 30 at SD9972. See PNCu (479); PNYW (7:211) and The DSL (SND), under wham.

whastle (E?) What is a whastle? It defies definition – no references can be found for it other than the showing on OSM and OSOL 16 at NT8010.

wheel (E?) The upper reaches of the River Bain in Yorks., are known as Low Wheel. Above Semer Water is High Wheel. Oliver (57) quotes a poem by Alexander Hume, a verse of which follows: *"The salmon out of cruives and creels, Up hauled into skouts; The bells and circles on the weills, Through louping of the trouts."* At the bottom of the page is a note stating that: "a *weil* or *wheel*, as the word is sometimes written, is a still, deep part of a river." Heslop (776) mentions this too. In view of the foregoing it would appear that 'wheel' is not a generic term but it is shown as such at NGR SD9186. In Bel., weel, USBG (Bel.:ivb), is used of a "pond" or "lake" which may be relative here.

wheel-drain (E) Loudon (710-11) best describes this agricultural drainage term: *"The wheel drain is a very ingenious invention, described in The Agricultural Report of the County of Essex. It consists of a draining-wheel of cast-iron, that weighs about 4 cwt. It is four feet in diameter; the cutting-edge or extremity of the circumference of the wheel is half an inch thick, and increases in thickness towards the centre. At fifteen inches deep it will cut a drain half an inch wide at the bottom, and four inches wide at the top."*

whelm (E) Found mainly in East Anglian counties, Young (1797:157) mentions it in answer to a question: "I strongly recommend these carrier ditches to be open, though at the expense of a whelm at the bottom of a field where a cart-way is necessary". Rye (245) gives: "Whelm. Half a hollow tree, placed with its hollow side downwards, to form a small watercourse." Moor (478-9) gives the same, and The EDD(6:426a) lists: "WELHAM", and the variant, "wellum", from a Suf. quotation. See the MED, under walm(e.

whip (E?) Just south of Hepple, in Northumberland, are two stream courses – one is a burn – the other, a whip. The burn offers no problem, but whip defies definition; Heslop's *Northumberland Words* draws a blank on this, so too, do other northern glossaries and PN volumes for the area, yet if burn is a course, then whip must surely be one, because it's shown as such on OSOL 42 where both courses can be seen centred on NY9998.

wielle (OE) See well.

wille (OE) See well.

wille-burne (OE) B-T (1228) give: *"running stream."* See walle- and well- stream and well.

wille-streám (OE) B-T (1228) give: *"running stream."* See walle- and well- stream and well.

willis (E) Elworthy (1889:108) defines willis as: "a rill from a spring" in a *Report of the Committee on Devonshire Verbal Provincialisms.* No other provincial or dialect dictionary lists the word other than The EDD(6.496b) which is referenced to the above.

winterbeck (E) See winterbourne.

winterbourne (E) From the OE *winterburna*, B-T (1236). This term is used of watercourses which are normally dry in summer and wet in winter; usually ones running in chalk landscapes. It enters into many OE charter names as *winterburna*, *winterburne* and *winterburninga* (BCS 226, 467, 667, 892, 1145, 1188, 1235, 1282 and 1286). From the *Lindisfarne Gospels*, (John 18:1) written about 950, Cook (208, under *winterburna*) gives the A-S form: "*uinterburna*". A number of Winterbourne PN's exist too; in Berks., Gloucs. Sx. and Wilts. There is a R Winterborne in Dor. and a R Winterburn in Yorks. A Winterbourne Stream can be seen north of Newbury at NGR SU4672, and, in addition, Berks has a Winterbrook, and Notts., a Winterbeck. See winter-burna, winterwell, B-T (747 & 1236) and the relative EPNS volumes. See winterwell.

winterbrook (E) See winterbourne.

winterburna (OE) See winterbourne.

winterburne (OE) See winterbourne.

winterburninga (OE) See winterbourne.

winter-trench (E) See trench-royal.

winterwell (E) A winterwell often refers to a spring that is 'dry in summer and wet in winter'. It is also used of a stream, much the same as winterbourne, the more common term, especially in OE charters. There is a winterwell shown at NGR SP0603 near Ampney in Gl. and winterwell names are shown on other maps – there is a Winterwell Lane in Chesterblade, Som., but here, probably to be associated with one of the many springs in the area. See PNGl (1.50, 4.188) and winterbourne.

wirli (SNn.) See orli.

*wise (OE) 'River' is suggested for this term by EPNE (2:270) and ERN (465-6).

wissing (ME) A somewhat rare ME term for 'a watercourse in the form of a pipe, conduit or channel for the purpose of supplying water to a particular place'. In the *Cursor Mundi*, a Northumbrian poem of the 14th C, Morris (1874:2: 685a, line

11942) gives: "wissing", from the Cotton and Göttingen Mss., and the alternative, "rennyng", from the Fairfax and Trinity Mss. The glossary (6:1789a) explains *"water wissing"* as *"watercourse"*. In the poem, 'wissing' brought water to the lake. The word is now obsolete; it has not come down to us in any form. See the MED, under wissing, for other forms.

wolf (E) Charnock (1880:53) gives this as: "an arch or culvert for water to pass through." It's really the Essex dialect word *hulve*, pronounced "oolve or woolve", according to Jepp (18) so wolf must have been another variant known to Charnock, the v in woolve here, has become f; in a number of dialects v is often substituted for f particularly in Dev., Som., Sx. and IOW words, but f for v is uncommon. Cope (99) in *Hampshire Words and Phrases* gives: "a water-vore", an instance of using v for f. See flot; fox; hulve; vlot; and especially PNEss (6 under Crouch). For a Dev. PN example, see Voss PND (254).

wood-drain (E) An agricultural drainage term, one of the many "-drain" prefix variants listed by Ogilvie (2.96a).

worli (SNn.) See orli.

wriggate (E) See riggot.

wring (E) Hill (11) gives: "Wring is the same as rhin, or hrin, or rhein, an open cut or drain. Hence, it is thought Wring-ton is the "town on the Wring," or rhin."

wrythe (E) See rithe.

wy (Co.) See guy.

wydraught (E) Peacock (1889:619) lists this word and states: "A gutter; a sewer (obsolete)", Phillips (1720) gives: "a watercourse, or water-passage", the same is in Kersey (1708), not surprising as Phillips was edited by Kersey. Willis (2.245) gives: "a Hows for the comyn wyddrowght", which is explained in the glossary (3.623) as follows: "The spelling here used gives the etymology: *wyddrought* is 'wide draught,' i.e. *chief* drain, *main* sewer." In view of the foregoing, it would appear that wydraught was used of a drain or sewer rather than a watercourse in spite of the earlier references in Phillips. As a simplex, draught is not disputed as a watercourse and is supported by map evidence.

wylle (OE) See well.

wylle-stream (E) See well- and wille- stream.

wysc (OGaul.) Chalmers (43) gives: "water, a stream, a river."

wysg (Wel.) Evans (2:886a) gives: "stream" and Richards (305b) defines this term as: "a current, a course, a stream." These definitions are not confirmed by the GPC. Cormac (65) states: "The esc here cited seems cognate with O. W. uisc now wysg 'a stream'." Massey (180), in his, *A Book of The Beginnings – Egyptian Water-Names*, has much to say about wysg: "The Welsh name of the river called *Esk* is the *Wysg*, and this points to a rapid or spreading water as the primary type of the rivers so named. The *Wysg* takes the English forms of *Guash* (compare *gush*) in Rutland, the *Washburn* in Yorkshire, and the *Wash*. *Washes* are outlets in the seashore, and in the fen country large spaces left at intervals between the river banks, for floods to expand in, are named *washes*. *Wash, Gwash, Gwysg* are represented in Egyptian by the word *Kash*, to *water, spread, be in flood, inundate*." There is a Nant Wysg in Wal., shown at NGR SN5555, exhibiting tautology. See ascaig; goush; uisg; wash; Bowen's map of Rutland, 1756, which gives: "guash or wash"; DPNW (484); ERN (465); PNRu (2) and the very detailed paper by Williams (1990) on the RN Wysg.

Y

yeo (E) This is the south-western form of OE *ea*. It's very common as a watercourse generic and as a RN in Dev. and Somerset.; another River Yeo can be found south of Bradford Abbas in Dorset at NGR ST5913. Bishop Hobhouse gives various ME forms of yeo, from the 1500's, in some of the Somerset *Church Wardens Accounts*, namely: "yew" (139, 157); "yowe" (168) and "yo" (170), the Glossary (241) explains yeo: "The main drain of a level." Most of the forms relate to watercourses, in some cases yeo can mean 'sluice' – which appears in the accounts (157) as "yere". Dev. has quite a number of Yeo RN's and much can be gleaned from PND

(17-8) and the many cross references cited. See ea and, for yeo as a generic, OSX 140 at ST3945 and OSX 154 at ST4471..

yk (Co.) See ick.

ysgŵd melin (Wel.) Evans (2:341b) gives this compound as: "millrace.

ystrym (Wel.) Evans (1:398b) gives: "watercourse" and Pughe & Pryse (2:636a) define this term as: "a main stream, a current, a channel; also called a rhin".

Z

zwi(j)n (Fle.) See swin.

SOURCES & BIBLIOGRAPHY

AD   *A Descriptive Catalogue of Ancient Deeds in the Public Record Office*, 6 Vols., HMSO, London, 1890-1915.

Adamnan   *Life of Saint Columba, Founder of Hy*, written by Adamnan, Reeves, William, (Ed.), Edmonston and Douglas, Edinburg, 1874.

Addy   Addy, Sidney Oldall, *A Glossary of Words used in the Neighbourhood of Sheffield*, EDS, Trübner & Co., London, 1888.

AI   *Itinerarivm Antonini Avgvsti*, Parthey, G. & Pinder, M., (Eds.), Berolini, Berlin, 1848.

ALI   *Ancient Laws of Ireland*, Hancock, W. Neilson, (Ed.), Vol 1; Richey, The Right Rev. Alexander George, (Ed.), Vol. 4, HMSO, A Thom & Co. and others, Dublin, Longmans & Co. and others, London, 1865 & 1879.

Allen   Allen, Thomas, *The History and Antiquities of London, Westminster, Southwark, and Parts Adjacent*, Vol. 3, London, 1828.

AND   *The Anglo-Norman Dictionary*, a project of Aberystwyth and Swansea Universities, eres., available from http://www.anglo-norman.net (accessed 8th November, 2014).

Angus   Angus, James Stout, *A Glossary of the Shetland Dialect*, Alexander Gardner, Paisley, 1914.

Annals of Ulster   *Annals of Ulster, Otherwise, Annals of Senat; A Chronicle of Irish Affairs from A.D. 431 to A.D. 1540*, 3 Vols., Hennessy, William M., (Ed., Vol. 1), McCarthy, B., (Ed., Vols. 2 & 3), HMSO, Dublin, 1887-1895.

AR (1771)   *The Annual Register*, printed for J. Dodsley, London, 1771.

AR (1786)   *The Annual Register*, 5th Edn., printed for J. Dodsley, London, 1786.

AR (1793)   *The Annual Register*, printed for J. Dodsley, London, 1793.

Archaeologia   *Archaeologia, or Miscellaneous Tracts relating to*

Antiquity, Society of Antiquaries, Vol. 37, London, 1857.

Arkell   Arkell, Thomas, *On the Drainage of Land*, pp 318-40, in The Journal of The Royal Agricultural Society of England, Vol. 4, John Murray, London, 1843.

Armstrong   Armstrong, R. A., *A Gaelic Dictionary*, Printed for James Duncan, London, 1825.

Artizan   *The Artizan, A Monthly Journal of the Operative Arts*, Simpkin, Marshall & Co., London, 1845.

ASC   *The Anglo-Saxon Chronicle according to the Several Original Authorities*, 2 Vols., Thorpe, B., (Ed.), Rolls Series, London, 1861.

Ashton   Ashton, John, *The Fleet, Its River, Prison and Marriages*, Scribner & Welford, New York, 1888.

Atkinson (1868)   Atkinson, The Rev. J. C., *A Glossary of the Cleveland Dialect*, John Russell Smith, London, 1868.

Atkinson (1891)   Atkinson, The Rev. J. C., *Forty Years in a oorland Parish*, Macmillan & Co., London & New York, 1891.

Axon   Axon, William E. A., (Ed.), *English Dialect Words of the Eighteenth Century as shown in the "Universal Etymological Dictionary" of Nathan Bailey*, Trübner & Co., London, 1883.

Ayenbite   Stevenson, The Rev. Joseph, (Ed.), *The Ayenbite of Inwyt*, J. B. Nicholls and Sons, London, 1855.

Bailey   Bailey, N, *The Universal Etymological English Dictionary*, 3rd Edn., printed for J Darby and others, London, 1726.

Baker (1843)   Baker, Robert, *Essex Draining* , p 36, in The Journal of The Royal Agricultural Society of England, Vol. 4, John Murray, London, 1843.

Baker (1854)   Baker, Anne Elizabeth, *Glossary of Northamptonshire Words and Phrases*, 2 Vols., John Russell Smith, London and Abel &

Sons; Mark Dorman, Northampton, 1854.

Balg — Balg, G. H., *A Comparative Glossary of the Gothic Language with especial reference to English and German,* Trubner & Co., and others, London, 1887-9.

Bannister (1871) — Bannister, The Rev., John, *A Glossary of Cornish Names,* Williams & Norgate, Edinburgh & London, J. R. Netherton, Truro, 1871.

Bannister (1916) — Bannister, A. T., *The Place-Names of Herefordshire,* Cambridge University Press, 1916.

Barbour — Barbour, John, *The Bruce,* edited by W. M. Mackenzie, Adam & Charles Black, London, 1909.

Baring — Baring, F. H., *Crundels,* pp 300-3 in, The English Historical Review, Vol. 24, Longmans, Green and Co., London, 1909.

Bartholomew — Bartholomew, J. G., (Ed.), *The Survey Gazetteer of the British Isles,* George Newnes, London, 1904.

Bates — Bates, Henry Walter, *The Naturalist on The Amazons,* George Routledge & Sons, London and E. P. Dutton & Co., New York, 1863.

Battle — *Descriptive Catalogue of the Original Charters, Royal Grants and Donations constituting the Muniments of Battle Abbey,* offered for sale by Thomas Thorpe, London, 1835.

BCAG — *The British Cultivator and Agricultural Review,* Vol. 1, Simpkin, Marshall and Co., London, 1844.

BCS — W. G. de Gray Birch (Ed.), *Cartularium Saxonicum,* 3 Vols., Whiting & Co., London, 1885-93.

Bede — *Venerabilis Baedae Historiam Ecclesiasticam,* 2 Vols., Charles Plummer, (Ed.), The Clarendon Press, Oxford, 1896.

Bedford — *Bedfordshire County Records, Notes and Extracts from the County Records,* Vol. 1, Compiled by Messrs Hardy & Page, C. F. Timæus, Printer, Bedford, 1907.

BFM   *The British Farmer's Magazine*. See Smith (1851).

Biggins   Biggins, Peter, *The Mill on the Floss map* available from http://www.communitywalk.com/the_mill_on_the_floss_map/map/465729 (accesed 08/11/2012).

Bisschop   Bisschop, E. V., *Hill's Flemish-English and English-Flemish Dictionary*, L.B. Hill, London, 1914.

BL   The British Library Online Gallery available from http:/www.bl.uk/onlinegallery (accessed to date).

Black   Black, Henry Campbell, *A Law Dictionary*, West Publishing Co., Minnesota, 1910.

Blackie   Blackie, C., *A Dictionary of Place-Names*, 3rd Edn., John Murray, London, 1887.

Blackmore   Blackmore, R. D., *Lorna Doone*, 20th Edn., Henry T. Coates, Philadelphia, 1882.

Blakeborough Blakeborough, Richard, *Dialect Glossary of Over 4,000 Words and Idioms now in use in the North Riding of Yorkshire*, W. Rapp & Sons Limited, Saltburn-by-the-Sea, 1912.

BM   Ellis, Henry J., & Bickley, Francis B., *Index to the Charters and Rolls in the Department of Manuscripts British Museum*, 2 Vols., Printed for The Trustees, sold by Longmans & Co. and others, London, (1900, Vol. 1) & (1912, Vol. 2).

Bock & Bruch Bock, Albert & Bruch, Benjamin, *An Outline of the Standard Written Form of Cornish*, Cornish Language Partnership, 2008.

Bond   Bond, Chris, *Selected Cornish Place-Name Elements*, available from http://cornish-place-names.wikidot.com/place-name-elements (accessed 18th May, 2011).

Borlase   Borlase, William, *Antiquities, Historical and Monumental, of the County of Cornwall*, W. Bowyer & J. Nichols, London, 1769.

Boswell   Boswell, George, *A Treatise on Watering Meadows*, 2nd Edn., Printed for J. Debrett, London, 1790.

Bosworth	*The Gothic and Anglo-Saxon Gospels in parallel columns with the versions of Wycliffe and Tyndale*, Bosworth, The Rev. Joseph, (Ed.), John Russell Smith, London, 1865.
Bowlker	*Bowlker's Art of Angling*, R. Jones, Ludlow & Longman, Brown & Co., London, 1854.
Bradley	Bradley, Henry, *Some Prehistoric River-Names*, in An English Miscellany, The Clarendon Press, Oxford 1901.
Brees	Brees, S. C., *A Glossary of Civil Engineering*, Tilt & Bogue, London, 1841.
Bret. L	Du Rusquec, H., *Nouveau Dictionnaire Pratique et Etymologique du Dialecte De Léon*, Ernest Leroux, Paris, 1895.
Bret. V	Ernault, Emile, *Dictionnaire Breton-Français du Dialecte de Vannes*, Lafolye Freres, Vannes, 1904.
Bridlington	Lancaster, W. T., (Ed.), *Abstracts of the Charters and other Documents contained in the Chartulary of the Priory of Bridlington*, privately published, Leeds, 1912.
Britten	Britten, J., *Old Country and Farming Words*, EDS, Trübner & Co. Limited, London, 1880.
Britton	Britton, John, *Wiltshire Words 1825*, in Reprinted Glossaries, Series B, Skeat, The Rev. W. W., (Ed.), EDS, Trübner & Co. Limited, London, 1879.
Brockett	Brockett, John Trotter, *A Glossary of North Country Words*, 3rd Edn., 2 Vols., Emerson Charnley, Bigg Market, Newcastle and Simpkin, Marshall and Co., London, 1846.
Broderick	Broderick, George, *A Dictionary of Manx Place-Names*, EPNS, Nottingham, 2006.
Brogden	Brogden, J. Ellett, *Provincial Words and Expressions Current in Lincolnshire*, Robert Hardwicke, London and The Gazette Office, Lincoln, 1866.
Broughton	Broughton, Rhoda, *Belinda*, Richard Bentley & Son, London 1887 (Gbooks – accessed 1st December, 2011).

B-T   Bosworth, The Rev. J. & Toller, T. Northcote, *An Anglo-Saxon Dictionary*, OUP, Oxford, 1898 and Toller, T. Northcote, *An Anglo-Saxon Dictionary: Supplement*, OUP, Oxford, 1921, and eres., hosted by the Faculty of Arts, Charles University, Prague, available from http://bosworth.ff.cuni.cz/ (accessed 2011 to date).

Buck   Buck, C. D., *A Dictionary of Selected Synonyms in the Principal Indo-European Languages*, University Of Chicago Press, 1949.

Budge   Budge, Sir E. A., Wallis, *An Egyptian Hieroglyphic Dictionary*, 2 Vols., John Murray, London, 1920.

Burke (1837)   Burke, John French, *British Husbandry*, Vol. 2, Baldwin & Craddock, London, 1837.

Burke (1841)   Burke, John French, *On Land Drainage, Subsoil-Ploughing and Irrigation*, John Murray, London, 1841.

CA   *Catholicon Anglicanum, an English-Latin Wordbook* dated 1483, Herrtage, Sidney J. H., (Ed.), EETS, Vol. 75, N. Trübner & Co., London, 1881.

Caesar   Caesar, Julius, Caesar's Gallic War (*De Bello Gallica*), James B. Greenough, Benjamin L. D'ooge and M. Grant Daniell, (Eds.), Ginn and Company, New York, 1898.

Callis   Callis, Robert, *The Reading of the famous and learned Robert Callis, Esq., upon the Statutes of Sewers, 23 Hen. VIII. c. 5.*, 4th Edn., William John Broderip (Ed.), printed for Joseph Butterworth & Son, London, 1824.

Camden   Camden, William, *Britannia*, 2nd Edn., 2 Vols., Gibson, E., (Ed.), London, 1722.

Carlyle   Carlyle, Thomas, *History of Frederick II of Prussia called Frederick The Great*, 8 Vols., Chapman & Hall, London, 1858-65.

Carr   Carr, W., *The Dialect of Craven in the West Riding of the County of York*, 2nd Edn., 2 Vols., Wm. Crofts, London and Robinson &

Hernaman, Leeds, 1828.

Catholicon — Lagadeuc, Jehan, *Le Catholicon*, (Facsimile of 1499 original), Le Men, René-François, (Ed.), Éditions et impression Corfmat, Lorient, 1867.

CDME — Mayhew, The Rev. A. L. & Skeat The Rev. Walter W., *A Concise Dictionary of Middle English*, The Clarendon Press, Oxford, 1888.

CEAJ — *The Civil Engineer and Architect's Journal for 1837-1838*, Published for the Proprietor, London 1838.

Chalmers — Chalmers, George, *Caledonia: A historical and Topographical Account of North Britain*, 2nd Edn., Vol. 1, Alexander Gardner, Paisley, 1887.

Chambers — Chambers, R. W., *Widsith, A Study in Old English Legend*, Cambridge University Press, 1912.

Charnock (1859) — Charnock, Richard Stephen, *Local Etymology*, Houlston & Wright, London, 1859.

Charnock (1870) — Charnock, Richard Stephen, *Patronymica Cornu-Britannica; or, The Etymology of Cornish Surnames*, Longmans, Green, Reader & Dyer, London, 1870.

Charnock (1880) — Charnock, Richard Stephen, *A Glossary of The Essex Dialect*, Trübner & Co., London, 1880.

Chaucer — *The Prologue to Chaucer's Canterbury Tales*, M'Leod, Walter, (Ed.), Longmans, Green & Co., London, 1871.

Chope — Chope, R. Pearse, *The Dialect of Hartland, Devonshire*, EDS, Kegan Paul, Trench, Trübner & Co., London, 1891.

CHR — Wright, Andrew, *Court Hand Restored*, 9th Edn., corrected and enlarged by Charles Trice Martin, London, 1879.

Clarke Hall — Clarke Hall, J. R., *A Concise Anglo-Saxon Dictionary*, 2nd Edn., The Macmillan Company, New York, 1916 and eres. available from http://www.gutenberg.org/ebooks/31543

	(accessed 21st January, 2012).
CL	*Celtic-Lexicon*, comprising an *English – Proto-Celtic Word List*, (EPC) a *Proto-Celtic – English Word List*, (PCE) and *An English-Old Irish Annotated Word List*, (EOI), available from www.wales.ac.uk, (accessed 9th December, 2008).
Clarke	Clarke, John Algernon, *On the Farming of Lincolnshire*, pp 259-414, in The Journal of The Royal Agricultural Society of England, Vol. 12, John Murray, London, 1851.
CLB	*The Coventry Leet Book or Mayor's Register*, Part 1, Mary Dormer Harris (Ed.), EETS, Kegan Paul, Trench, Trübner & Co. Limited, London, 1907.
CLO	*The Carlyle Letters Online* (CLO) 2007, available from http://carlyleletters.org (accessed 3rd December, 2011).
CM	*The Celtic magazine*, Vol. 9, A & W MacKenzie, Inverness, 1884.
CodexA	*Codex Argenteus,* available from http://www.ub.uu.se/collections/selections-of-special-items-and-collections/silver-bible (accessed 16th January, 2012).
Coe	Coe, Jonathan Baron, *The Place-Names of the Book of Llandaf*, PhD Thesis, University of Wales, Aberystwyth, 2001.
COED	*The Compact Edition Of The Oxford English Dictionary*, Book Club Associates, London, 1979.
Cole	Cole, The Rev. R. E. G., *A Glossary of Words used in South-West Lincolnshire (Wapentake of Graffoe)*, EDS, Trübner & Co., London, 1886.
Coneys	Coneys, Thomas de Vere, *Focloir Gaoidhlige-Sacs-Beurla or An Irish-English Dictionary*, Hodges & Smith, Dublin, 1849.
Conway	Conway, Agnes E., *The Owners of Allington Castle, Maidstone (1086-1279)* in Archæologia Cantiana, Vol. 29, Transactions of the Kent

Archæological Society, London, 1911.

Cope   Cope, The Rev. Sir William H., *A Glossary of Hampshire Words and Phrases*, EDS, Trübner & Co., London, 1883.

Cormac   *Cormac's Glossary*, O'Donovan, John, (Ed.), The Irish Archaeological and Celtic Society, Calcutta, 1868.

Cornhill   *The Cornhill Magazine*, Vol. 62, Smith, George, (Founder), London, 1890 (Gbooks – accessed 10th January, 2012).

Cornish   Cornish, C. J., *The Naturalist on The Thames*, Seeley & Co., Ltd., London, 1902.

Cotgrave   Cotgrave, Randle, *Dictionarie of the French and English Tongues*, Adam Islip (Printer), London, 1611.

Courtney & Couch   Courtney, Miss M. A. & Couch, Thomas Q., *Glossary of Words in use in Cornwall – West Cornwall, Miss Courtney; East Cornwall, Thomas Couch*, EDS, Trübner & Co., London 1880.

CPNE   Padel, O. J., *Cornish Place-Name Elements*, 2 Vols., EPNS, 56-57, Cambridge University Press, 1985.

CPNS   Watson, W. J., *The History Of The Celtic Place-Names Of Scotland*, Edinburgh and London, 1926, New Edn., with Introduction by Simon Taylor, Birlinn Ltd, Edinburgh, 2004.

CR   *The Celtic Review*, Vol. 10, December 1914 To June 1916, William Hodge & Co., Edinburgh; David Nutt, London, and Hodges, Figgis & Co. Ltd, Dublin, 1916.

Cregeen   Cregeen, Archibald, *A Dictionary of the Manks Language*, J. Quiggin, Douglas; Whittaker, Treacher & Arnot, London and Evans, Chegwin & Hall, Liverpool, 1835.

Crossing   Crossing, William, *Guide to Dartmoor: A Topographical Description of the Forest and Commons*, 3rd Edn., 2 Vols., A. Wheaton & Co., Ltd, Exeter, 1914.

CSP   Camden Society Publications.

Cullum   Cullum, The Rev. Sir John, *Suffolk Words, 1813,* in Reprinted Glossaries, Series B, Skeat, The Rev. W. W., EDS, (Ed.), Trübner & Co. Limited, London, 1879.

Cunliffe   Cunliffe, Henry, *A Glossary of Rochdale-with-Rossendale Words and Phrases*, J. Heywood, Manchester, 1886.

Darlington   Darlington, Thomas, *The Folk-Speech of South Cheshire*, EDS, Trübner & Co. Limited, London, 1887.

Dartmoor   Available from http://www.legendary dartmoor.co.uk/index.htm, (accessed 2nd August, 2013).

Dartnell   Dartnell, George Edward and Goddard, The Rev. Edward Hungerford, *A Glossary of Words used in the County of Wiltshire*, EDS, Henry Frowde, London 1893.

Darton   Darton, M., *The Dictionary Of Scottish Place-Names*, Lochar Publishing, Moffat, 1990.

Darwin   Darwin, Erasmus, *The Botanic Garden*, Jones & Company, London, 1824.

Davies   Davies, The Rev. John, *The Celtic Element in the Dialectic Words of the Counties of Northampton & Leicester*, reprinted from Archaeologia Cambrensis, 5th Series, Vol. 2, No. 5, London, 1885.

Davis   Davis, Thomas, *A Report of the State of Agriculture in the County of Wiltshire*, The Board of Agriculture, printed for Richard Phillips, London, 1811.

DBO   *Origins of Place-Names* in the *Domesday Book Online* available from http://www.domesdaybook.co.uk/places.h tml (accessed 25th September, 2010).

Defoe   Defoe, Daniel, *A Tour thro' the Whole Island of Great Britain*, 6th Edn., Vol. 2, printed for D. Browne and others, London, 1761.

DEPN   Ekwall, E., *The Concise Oxford Dictionary Of English Place-Names*, 3rd Edn., Oxford University Press, 1947.

Dexter (1871)   Dexter, T. F. G., *A Glossary of Cornish Names*,

	Williams & Norgate, London and J. R. Netherton, Truro, 1871.
Dexter (1926)	Dexter, T. F. G., *Cornish Names*, Longmans, Green & Co Ltd, London, 1926.
Dickinson	Dickinson, W., *A Glossary of the words and phrases pertaining to the Dialect of Cumberland*, re-arranged, illustrated and augmented with quotations by E. W. Prevost, Bemrose and Sons, London; Thurnam and Son, Carlisle, 1899.
Dickson (1765)	Dickson, Adam, *A Treatise of Agriculture*, 2nd Edn., A. Kincaid and J. Bell, Edinburgh, 1765.
Dickson (1805-7)	Dickson, R. W., *Practical Agriculture, or A Complete System of Modern Husbandry*, 2 Vols., Vol.1 1805 and 2nd Edn., 1807; Vol. 2 1807, Richard Phillips, London, 1805-7.
Dineen (1900)	Dinneen, Patrick S., *Dánta Aodhagáin Uí Rathaille, The Poems of Egan O'Rahilly*, published for The Irish Texts Society, David Nutt, London, 1900.
Dineen (1904)	Dinneen, Patrick S., *Foclóir Gaedhilge agus Béarla, An Irish-English Dictionary*, published for The Irish Texts Society, M. H. Gill & Son, Ltd and The Gaelic League, Dublin; David Nutt, London, 1904.
Dinsdale	Dinsdale, Frederick P., *A Glossary of Provincial Words used in Teesdale in the County of Durham*, J. R. Smith & George Bell, London; John Atkinson, Barnard Castle; Matthew Bell & T. & A. Bowman, Richmond, 1849.
DOST	See DSL.
Dottin	Dottin, Georges, *La Langue Gauloise*, Libairie C. Klincksiecek, Paris, 1918.
DPNW	Owen, Hywel Wyn and Morgan, Richard, *Dictionary of the Place-Names of Wales*, Gomer Press, Llandysul, 2007, reprinted 2008.
Drake	Drake, Nathan, *A Journal of The First and Second Sieges of Pontefract Castle*, 1644-5, p xvii, in Surtees Society Publications, Vol. 37,

	George Andrews, Durham, and others, 1860.
Drayton	*The Complete Works Of Michael Drayton*, now first collected with introductions and notes by The Rev. Richard Hooper, Vols. 1-3, *Polyolbion*, John Russell Smith, London, 1876.
DSC	*Dictionarium Scoto-Celticum: A Dictionary of the Gaelic Language*, Compiled by The Higland Society of Scotland, 2 Vols., William Blackwood, Edinburgh and T Cadell, London, 1828.
DSL	*The Dictionary of the Scots Language* (DSL) comprises electronic editions of the two major dictionaries of the Scots Language: the 12-Volume *Dictionary Of The Older Scottish Tongue* (DOST) and the 10-Volume *Scottish National Dictionary* (SND), eres., available from http://www.dsl.ac.uk/ (accessed 2005 to date).
Du Cange	Du Cange, Charles Dufresne, *Glossarium mediæ et infimæ latinitatis*. Niort : L. Favre, 1883-1887, searchable full-text online edition, by the École nationale des chartes available from http://ducange.enc.sorbonne.fr/ (accessed 2013 to date).
Duignan (1902)	Duignan, W. H., *Notes On Staffordshire Place-Names*, Henry Frowde, London, 1902.
Duignan (1905)	Duignan, W. H., *Worcestershire Place-Names*, Henry Frowde, London, 1905.
Duignan (1912)	Duignan, W. H., *Warwickshire Place-Names*, Henry Frowde, London, 1912.
Duncumb	Duncumb, John, *General View of the Agriculture of the County of Hereford*, Board of Agriculture, W Bulmer & Co., (Printer) London, 1805.
Durham Roll	*Extracts from the Account Rolls of the Abbey of Durham*, 3 Vols., Fowler, J. T., (Ed.), Surtees Society Publications, Vols. 99, 100 & 103, Andrews & Co., Durham, and others, 1898-1901.

Dwelly	Dwelly, E., *The Illustrated Gaelic-English Dictionary*, 3 Vols., Author publication, Fleet, 1918.
Earle (1865)	Earle, John, (Ed.), *Two of the Saxon Chronicles Parallel with Supplementary Extracts from the Others*, The Clarendon Press, Oxford, 1865.
Earle (1888)	Earle, John, *A hand Book of Land Charters, and other Saxonic Documents*, The Clarendon Press, Oxford, 1888.
Earls	*The Flight of the Earls*, Ó Cianáin, Tadhg & Walsh, The Rev. Paul (Ed.), M. H. Gill & Son Ltd, Dublin, 1916.
Easther	Easther, The Rev. Alfred, & Lees The Rev.Thomas, (Ed.), *A Glossary of the Dialect of Almondbury and Huddersfield*, EDS, Trübner & Co., London, 1883.
EDD	*The English Dialect Dictionary*, 6 Vols., Wright, Joseph, (Ed.), Henry Frowde, Oxford & London and G. P. Putnam's Sons, New York, 1898-1905.
eDIL	*electronic Dictionary of the Irish Language*, eres., available from http://www.dil.ie/ (accessed 2006 to date).
Edmondston	Edmondston, Thomas, *An Etymological Glossary of the Shetland & Orkney Dialect*, Adam and Charles Black, Edinburgh, 1866.
EDS	English Dialect Society Publications.
Edwards	Edwards, Edward, (Ed.), *Liber Monasterii De Hyda*, Longmans, Green, Reader, and Dyer. London, 1866.
EETS	Early English Text Society Publications.
Ellis	Ellis, William, *The Modern Husbandman, or the Practice of Farming*, 4 Vols., T. Osbourne and M. Cooper, London, 1744.
Ellwood (1894)	Ellwood, The Rev. T., *The Landnama Book of Iceland*, T. Wilson (Printer), Kendal, 1894.
Ellwood (1895)	Ellwood, The Rev. T., *Lakeland and Iceland being a Glossary of Words in the Dialect of Cumberland, Westmorland and North Lancashire*, EDS, Henry Frowde, London, 1895.

Elworthy (1886)   Elworthy, F. T., *The West Somerset Word-Book*, EDS, Trübner & Co. Limited, London, 1886.

Elworthy (1889)   Elworthy, F. T., *Eleventh Report of the Committee on Devonshire Verbal Provincialisms*, p 108, in Reports & Transactions of the Devonshire Association, Vol. 21, W. Brendon & Son, Plymouth, 1889.

Embleton   Embleton, D., *A Catalogue of Place-Names in Teesdale*, in Natural History Transactions of Northumberland, Vol. 9, Durham and Newcastle-On-Tyne, Williams & Norgate, London and F & W Dodsworth, Newcastle, 1888.

EPNE   Smith, A. H., *English Place-Name Elements*, 2 Vols., EPNS 25-26, Cambridge University Press, 1956.

EPNS   English Place-Name Society Publications   .

ERN   Ekwall, E., *English River-Names*, The Clarendon Press, Oxford, 1928.

Evans (1852/8)   Evans, Daniel Silvan, *An English and Welsh Dictionary*, 2 Vols., Robert Gee, Denbigh and Simpkin & Marshall, London, Vol. 1, 1852 & Vol. 2, 1858.

Evans (1881)   Evans, Arthur Benoni, *Leicestershire Words, Phrases and Proverbs*, Sebastian Evans (Ed.), Trübner & Co., London, 1881.

Evans (1906)   Evans, Evans, J. G., (Ed.), *The Black Book of Carmarthen*, Pwllheli, 1906.

Eyton   Eyton, The Rev. R. W., *Antiquities of Shropshire*, Vol. 9, John Russell Smith, London 1859.

Farmer   Farmer, A., *Place-Name Synonyms Classified*, David Nutt, London, 1904.

FD   *How a Water Meadow Works*, eres., available from www.farm-direct.co.uk/farming /history/watermeadow/how.html (accessed 25th August, 2014).

Fees   *Liber Feodorum, The Book of Fees, commonly called Testa de Nevill*, Part 1, A. D. 1198-1242, HMSO, London, 1920.

Fenland N&Q | *Fenland Notes & Queries*, Vol. 1, Saunders, W. H. Bernard, (Ed.), Geo. G. Caster, Peterborough, 1891.

Fick, Falk & Torp | Fick, A., & Falk, H., (Ed. by Torp, A., 1909) *Wörterbuch der Indogermanischen Sprachen*, eres., created by Sean Crist, available from http://lexicon.ff.cuni.cz/texts/pgmc_torp_about.html (accessed 25th November 2011).

Field (1972) | Field, J., *English Field Names*, David & Charles, Newton Abbot, 1972.

Field (1993) | Field, J., *A History of English Field Names*, Longman, London, 1993.

Finchale | *The Priory of Finchale; The Charters of Endowment, Inventories and Account Rolls of The priory of Finchale in the County of Durham*, Raine, James, (Ed.), Surtees Society Publications, Vol. 6, J. B. Nicholls & Son and William Pickering, London, 1837.

Flanagan | Flanagan, D. & L., *Irish Place Names*, Gill & Macmillan, Dublin, 2002.

Forbiger | Forbiger, Albert, *Handbuch der Alten Geographie von Europa*, Hamburg, 1877.

Forby | Forby, The Rev Robert, *The Vocabulary of East Anglia*, 2 Vols., J. B. Nicholls and Sons, London, 1830.

Forni | Forni, Gianfranco, *A First Etymological Dictionary of Basque as an Indo-European Language*, Amazon.co.uk, Ltd., (Printer), 2014.

Forsberg | Forsberg, R., *A Contribution To A Dictionary Of Old English Place-Names*, Nomina Germanica No. 9, Almqvist & Wiksells, Uppsala, 1950.

Forward | Forward, Eleanor J., *Place-names of the Whittlewood area*, Ph.D. thesis, University of Nottingham, 2007, eres., available from etheses.nottingham.ac.uk/568/1/Thesis_F.F.pdf (accessed 30th November, 2010).

Furley | Furley, Robert, *An Outline of the History of Romney Marsh*, pp 178-200, in Archæologia Cantiana, Vol. 13, Transactions of the Kent

Archæological Society, London, 1880.

Furnivall   Furnival, Frederick J., *Early English Poems and Lives of Saints*, Published for The Philological Society, A. Asher & Co., Berlin, 1862.

Galway L   Galway Library, Placenames of Galway database available from http://places. galwaylibrary.ie/ (accessed 18th May, 2011).

Geograph   The Geograph Britain and Ireland project aims to collect geographically representative photographs and information for every square kilometre of Great Britain and Ireland, using the National Grid References (NGR). An excellent resource for viewing watercourses nationwide, available from www.geograph.org.uk (accessed to date).

George (1998)   George, Dr. K. J, (Ed.), *Gerlyver Kernewek Kemmyn - An Gerlyver Kres, The New Standard Cornish Dictionary, Cornish-English, English-Cornish Dictionary*, Cornish Language Board, 1998.

George (2009)   George, Dr. K. J, (Ed.), *An Gerlyver Meur: Cornish-English, English-Cornish Dictionary*, 2nd Edn., Cornish Language Board, 2009.

Gepp   Gepp, Edward, *A Contribution to An Essex Dialect Dictionary*, George Routledge & Sons, London, 1920.

Gilchrist   Gilchrist, R. Murray, *The Peak District*, Blackie & Son Ltd, London and Glasgow, undated.

GKN   *Gerlever Kernûak Nowedga* (a Modern Cornish Dictionary) available from http://home.btconnect.com/htm_cornwall / (accessed 5th May 2013).

Glamorgan   *Cartæ et Alia Munimenta quæ ad Dominium de Glamorgan*, 6 Vols., William Lewis (Printer), Cardiff, 1885-1910.

Glasscock   Glasscock, Jun., J. L., (Ed.), *The Records of St. Michael's Parish Church Bishop's Stortford*, Elliot Stock, London and A. Boardman, Bishop's Stortford, 1882.

GN — *Geographic Names* available from http://www. geographic. org/ geographic_names /ireland/o/ow.html (accessed 25th May, 2012).

Godstow — *The English Register of Godstow Nunnery, near Oxford, written about 1450*, Clarke, Andrew, (Ed.), EETS, Kegan Paul, Trench, Trübner & Co. Limited and Henry Frowde, London, 1911.

Goodall — Goodall, Armitage, *Place-Names of South-West Yorkshire*, The University Press, Cambridge, 1914.

Gower (1876) — Gower, Granville Leveson, *Surrey Provincialisms*, EDS, Trübner & Co., London, 1876.

Gower (1893) — Gower, Granville Leveson, *A Glossary of Surrey Words*, EDS, Henry Frowde, London 1893.

GPC — *Geiriadur Prifysgol Cymru* , Thomas, R. J., Bevan, Gareth A. & Donovan, P. J. (Eds.) The University of Wales Dictionary, 4 Vols., University Of Wales Press, Cardiff, 1967-2002 (2nd Edn. in progress 2003-) and eres., available from http://www.cymru.ac.uk/geiriadur/gpc_pd fs.htm (accessed 2006 to date).

Greene — Greene, Robert, *The Dramatic Works of Robert Greene*, to which are added his poems with some account of the author, and notes by The Rev. Alexander Dyce, Vol. 2, William Pickering, London, 1831.

Gregor — Gregor, The Rev. Walter, *The Dialect of Banffshire with a Glossary of Words not in Jamieson's Scottish Dictionary*, Asher and Co., London, 1866.

Grimm — Grimm, Conrad, *Glossar zum Vespasian-Psalter und den Hymnen*, Carl Winter's Universitätsbuchhandlung, Heidelburg, 1906.

Grose — Grose, Francis, *A Glossary of Provincial and Local Words used in England*, (with supplement by Samuel Pegge), John Russell Smith, London,

1839.

Grundy   Grundy, G. B., *On the Meaning of Certain Terms in the Anglo-Saxon Charters*, pp 47-50, in Essays and Studies by members of The English Association, Vol. 8, The Clarendon Press, Oxford, 1922.

Gwilt   Gwilt, Joseph, *An Encyclopædia of Architecture, Historical, Theoretical and Practical*, Longman, Brown, Green & Longmans, London, 1842.

Halliwell   Halliwell, J. O., *A Dictionary of Archaic and Provincial Words*, Routledge & Sons Ltd, London, 1904.

Halloran   Halloran, John A., *Sumerian Lexicon*, Version 3.0, available from www.sumerian.org/sumerian.pdf (accessed 2004 to date).

Harland   Harland, Captain John, *A Glossary of Words used in Swaledale, Yorkshire*, EDS, Trübner & Co., London, 1873.

Harrison (1877)   *Harrison's Description of England in Shakspere's Youth, Part 1, The Second Book*, Furnivall, Frederick J. (Ed.), Trübner & Co., London, 1877.

Harrison (1898)   Harrison, Henry, *The Place-Names of the Liverpool District*, Elliot Stock, London, 1898.

Hartshorne   Hartshorne, The Rev. Charles Henry, *Salopia Antiqua*, John W. Parker, London, 1841.

Harvey   Harvey, The Rev. E. G., *Mullyon: Its History, Scenery and Antiquities*, W. Lake, Truro and Simpkin, Marshall & Co., London, 1875.

Hazlitt   Hazlitt, W. Carew, (Ed.), *The Whole Works of William Browne*, Vol. 1, Roxburghe Library, London, 1868.

Hennessy   Hennessy, William M., (Ed.), *Chronicum Scotorum. A Chronicle of Irish Affairs, from the Earliest Times to A.D.1135*, Longmans, Green, Reader and Dyer, London, 1860.

Henry   Henry, Victor, *Lexique Étymologique des termes les plus usuels du Breton Moderne*, J. Plihon et L. Hervé, Rennes, 1900.

Henryson	Henryson, Robert, *The Poems and Fables of Robert Henryson, now first collected with notes and a memoir of his life by David Laing*, William Patterson, Edinburgh, 1865.
Heslop	Heslop, The Rev. Oliver, *Northumberland Words. A Glossary of Words used in the County of Northumberland and on the Tyneside*, 2 Vols., Vol. 1, EDS, Kegan Paul, Trench, Trübner & Co. Limited, London, 1892, Vol. 2, EDS, Henry Frowde, London, 1893-4.
Hessels (1890)	Hessels, J. H., (Ed.), *An Eighth-Century Latin-Anglo-Saxon Glossary, Preserved in the Library of Corpus Christi College, Cambridge*, The University Press, Cambridge, 1890.
Hessels (1906)	Hessels, J. H., (Ed.), *An Eighth-Century Latin-Anglo-Saxon Glossary, Preserved in the Library of Leiden University*, The University Press, Cambridge, 1906.
Hessens	Hessens, H., *Irisches Lexikon*, Halle, 1933.
Hill	Hill, James S., *The Place-Names Of Somerset*, St. Stephen's Printing Works, Bristol, 1914.
Hobhouse	Hobhouse, The Right Rev. Bishop, (Ed.), *Church Wardens Accounts of Croscombe, Pilton, Yatton, Tintinhull, Morebath and St. Michael's, Bath, Ranging from AD 1349 to 1560*, Somerset Record Society, Harrison & Sons (Printer), London, 1890.
Hobkirk	Hobkirk, Chas. P., *Huddersfield: Its History and Natural History*, 2nd. Edn., Geo. Tindall, Huddersfield and Simpkin, Marshall & Co., London, 1868.
Hogan	Hogan, Edmund, *Onomasticon Goedelicum*, Hodges, Figgis & Co., Limited and Williams & Norgate, London, 1910.
Holder	Holder, A., *Alt-Celtischer Sprachschatz*, Teubner, Liepzig, 1896.
Holinshed	Holinshed Raphael, *Holinshed's Chronicles of England, Scotland and Ireland*, 6 Vols., printed for J. Johnson & others, London, 1807.
Holland	Holland, Robert, *A Glossary of Words used in the*

	County of Chester, Trübner & Co., London, 1885.
Hooke	Hooke, Della, *Worcestershire Anglo-Saxon Charter Bounds*, Boydell Press, Suffolk, 1990.
Horovitz	Horovitz, David, *A Survey and Analysis of the Place-Names of Staffordshire*, Thesis submitted to the University of Nottingham for the degree of Doctor of Philosophy, October 2003.
Hutchinson	Hutchinson, William, *The History of the County of Cumberland*, Vol. 2, E Jollie (Printer), Carlisle, 1794.
IED	*An Icelandic-English Dictionary*, based on the ms. collections of the late Richard Cleasby, enlarged and completed by Gudbrand Vigfusson, The Clarendon Press, Oxford, 1874.
Inchaffray	*Charters Bulls and other documents relating to The - Abbey Of Inchaffray*, William Alexander Lindsay and others (Eds.), The University Press, Edinburgh, 1908.
Index (B)	Basden, E. B., (Compiler), *Index of Celtic Elements in Professor W J Watson's The History Of The Celtic Place-Names Of Scotland*, Scottish Place-Name Society, Edinburgh, 1997.
Index (E)	James, A. G., & Taylor, S., *Index of Celtic and Other Elements in W.J.Watson's The History of the Celtic Place-names of Scotland* incorporating the work of A. Watson and the late E. J. Basden, Scottish Place-Name Society, Edinburgh, eres., available from http://www.spns.org.uk/WatsIndex2.html (accessed 30th September, 2010).
IOMS	*Isle of Man 1/25,000 Outdoor Leisure Map*, 2 Sheets, North & South, 3rd Digital Edition, Isle of Man Survey, Douglas, 2009.
J & N	*Jim and Nell: A dramatic Poem in the Dialect of North Devon*, by a Devonshire Man, Unwin Brothers (Printer), London, 1867.
Jackson	Jackson, Georgina, *Shropshire Word-Book, A*

	Glossary of Archaic and Provincial Words, etc, *used in the County,* Trübner & Co. Limited, London; Adnitt & Naunton, Shrewsbury and Minshull & Hughes, Chester, 1879.
Jago (1882)	Jago, Fred. W. P., *The Ancient Language and the Dialect of Cornwall,* Netherton & Worth, Truro, 1882.
Jago (1887)	Jago, Fred. W. P., *An English-Cornish Dictionary,* Simpkin, Marshall & Co., London and W. H. Luke, Plymouth, 1887.
Jakobsen (1897)	Jakobsen, Jacob, *The Dialect and Place-Names of Shetland,* T. & J. Manson, Lerwick, 1897.
Jakobsen (1928)	Jakobsen, Jakob, *An Etymological Dictionary of the Norn Language in Shetland,* 2 Vols., David Nutt, London and Vilhelm Prior, Copenhagen, 1928.
James	James, Alan, *Brittonic Language in the Old North,* (2001-) Scottish Place-Name Society, Edinburgh, eres., available from http://www.spns.org.uk/bliton/blurb.html (accessed 30th September, 2010).
Jamieson	Jamieson, John. (Longmuir, J. & Donaldson, D., Eds.) *An Etymological Dictionary of the Scottish Language,* 5 Vols., Alexander Gardner, Paisley, 1879-87.
Jellinghaus	Jellinghaus, H, *Die Westfälischen Ortsnamen,* Lipsius & Tischer, Kiel and Leipzig, 1896.
Jennings (1869)	Jennings, James, *The Dialect of the West of England particularly Somersetshire,* 2nd Edn., John Russell Smith, London, 1869.
Jennings (1907)	Jennings, Louis J., *Field Paths and Green Lanes in Surrey and Sussex,* 5th Edn., John Murray, London, 1907.
JEPNS	Journal of the English Place-Name Society.
Jerome	Jerome, K. Jerome, *Three Men in a Boat,* Henry Holt and Company, New York, 1890.
Johnston	Johnston, James B., *Place-Names of Scotland,* David Douglas, Edinburgh, 1892.
Johnstone	Johnstone, John, *An Account of the Mode of Draining Land, according to the system practised*

by Mr. *Joseph Elkington*, The Board of
Agriculture, London, 1801.

Joyce (1869)   Joyce, P. W., *The Origin and History of Irish
Names of Places*, 3 Vols., Phoenix Publishing
Co., Ltd., Dublin, Cork and Belfast, 1869-
1913.

Joyce (1922)   Joyce, P. W., *Irish Local Names Explained*, The
Educational Company of Ireland Limited,
Dublin and Longmans, Green & Company,
London, 1922.

Joyce (1984)   Joyce, P. W., *Pocket Guide To Irish Place-Names*,
The Appletree Press Ltd, Belfast, 1984.

KCD   Kemble, J. M., *Codex Diplomaticus Aevi Saxonici*,
6 Vols., English Historical Society, London
1839-48.

Kelham   Kelham, Robert, *A Dictionary of the Norman or
Old French Language*, printed for Edward
Brooke, Successor to Messrs Worrall &
Tovey, London, 1779.

Kelly   Kelly, John, (Gill, W & Clarke, J. T, Eds.)
*Fockleyr Manninagh as Baarlagh, The Manx
Dictionary in Two Parts: First, Manx & English;
and the second, English & Manx*, Manx Society
Publications, Vol. 13, Douglas, 1866.

Kennedy   Kennedy, James. *Glenochel; A Descriptive Poem*,
Vol. 1, R Chapman (Printer), Glasgow and
London, 1810.

Kennett   Kennett, White, *Parochial Antiquities,1695*, in
Reprinted Glossaries, Series B, Skeat, The
Rev. W. W., (Ed.), EDS, Trübner & Co.
Limited, London, 1879.

Kersey   Kersey, John, *Dictionarium Anglo-Britannicum or
A General English Dictionary*, London, 1708.

Kiliani   Kiliani, Cornelii, *Etymologicvm Tevtonicæ Lingvæ:
sive Dictionarivm Tevtonico-Latinvm*, Officina
Plantiniana, Antverpiæ, 1599, (original) and
ebook, available from
http://www.dbnl.org/tekst/kili001etym01_
01/downloads.php (accessed 24th
November, 2011).

King   King, Jacob, *Analytical Tools for Toponymy: Their Application to Scottish Hydronymy*. Ph.D. thesis, University of Edinburgh, 2008, eres., available from http://hdl.handle.net/1842/3020 (accessed 30th May, 2010).

Kirby   Kirby, B., *Lakeland Words*, T. Wilson (Printer), Kendal, 1898.

Klussis   Klussis, M., *Dictionary of Revived Prussian: Prussian-English English-Prussian*, eres., available from http://donelaitis.vdu.lt/prussian/Engl.pdf (accessed 25th November 2011).

Kneen   Kneen, J. J., *The Place-Names of the Isle of Man with their Origin and History*, The Manx Society, Douglas, 1925.

Knight   Knight, Edward H., *Knight's American Mechanical Dictionary*, J. B. Ford and Company, New York, 1874.

Köbler (Gm.)   Köbler, G., *Germanisches Wörterbuch*, 5th Edn., 2014, eres., available from http://www.koeblergerhard.de/publikat.html (accessed January 30th, 2015 to date).

Köbler (Got.)   Köbler, G., *Gotisches Wörterbuch*, 4th Edn., 2014, eres., available from http://www.koeblergerhard.de/publikat.html (accessed January 30th, 2015 to date).

Köbler (Got.)   Köbler, G., *Gotisches Wörterbuch*, 4th Edn., 2014, eres., available from http://www.koeblergerhard.de/publikat.html (accessed January 30th, 2015 to date).

Köbler (IG)   Köbler, G., *Indogermanisches Wörterbuch*, 5th Edn., 2014, eres., available from http://www.koeblergerhard.de/publikat.html (accessed January 30th, 2015 to date).

Köbler (OE)   Köbler, G., *Altenglisches Wörterbuch*, 4th Edn., 2014, eres., available from http://www.koeblergerhard.de/publikat.html (accessed January 30th, 2015 to date).

Köbler (OFris.)   Köbler, G., *Altfriesisches Wörterbuch*, 4th Edn.,

2014, eres., available from
http://www.koeblergerhard.de/publikat.ht
ml (accessed January 30th, 2015 to date).

Köbler (OHG)   Köbler, G., *Althochdeutsches Wörterbuch*, 6th
Edn., 2014, eres., available from
http://www.koeblergerhard.de/publikat.ht
ml (accessed January 30th, 2015 to date).

Köbler (OLFrk.)   Köbler, G., *Altniederfränkisches Wörterbuch*, 5th
Edn., 2014, eres., available from
http://www.koeblergerhard.de/publikat.ht
ml (accessed January 30th, 2015 to date)

Köbler (ON)   Köbler, G., *Altnordisches Wörterbuch*, 4th Edn,
2014, eres., available from
http://www.koeblergerhard.de/publikat.ht
ml (accessed January 30th, 2015 to date).

Köbler (OSax.)   Köbler, G., *Altsächsisches Wörterbuch*, 5th Edn.,
2014, eres., available from
http://www.koeblergerhard.de/publikat.ht
ml (accessed January 30th, 2015 to date).

LC   *LacusCurtius*, for all things Roman, available
from
http://penelope.uchicago.edu/Thayer/E/
home.html (accessed January 30th, 2015).

L & S   Lewis, L. T., & Short, C., *A Latin Dictionary*,
The Clarendon Press, Oxford, and Harper
& Brothers, New York, 1879 and eres.,
available from
http://www.inrebus.com/latindictionary.p
hp (accessed to date).

Lambarde   Lambarde, William, *A Perambulation of Kent:
Conteining the Description, Hystorie and Customes
of That Shire. Written in the Yeere 1570, by
William Lambarde*, Baldwin, Cradock & Joy,
London, 1826.

Langtoft   *The Chronicle of Pierre De Langtoft*, Vol. 2, Wright,
Thomas, (Ed.), Longmans, Green, Reader,
and Dyer, London, 1868.

Lawson (1884a)   Lawson, Emily M., *The Nation in the Parish or
Records of Upton-On-Severn*, Houghton &
Gunn, London & W. E. Cooper, Upton-

On-Severn, 1884.

Lawson (1884b) — Lawson, Robert, *Upton-On-Severn Words and Phrases*, EDS, Trübner & Co., London, 1884.

Layamon — *Layamon's Brut or Chronicle of Britain*, Sir Frederic Madden (Ed.), 3 Vols., The Society Of Antiquaries, London, 1847.

Le Gonidec — Le Gonidec, Jean-François, *Dictionnaire Breton-Français*, Vol. 2, L Prud'homme, Saint-Brieuc, 1850.

Lediard — Lediard, Thomas, *The Life of John, Duke of Marlborough*, 2nd Edn., Vol.1, J. Wilcox, London, 1743.

Leigh — Leigh, Lieutenant-Colonel Egerton, *A Glossary of Words used in The Dialect of Cheshire*, founded on a similar attempt by Roger Wilbraham, Hamilton, Adams & Co., London; Minshull & Hughes, Chester, 1877.

Leland — *The Itinerary Of John Leland In Or About The Years 1535-1543*, 11 Parts in 5 Vols., Lucy Toulmin Smith, (Ed.), George Bell & Sons, London, 1906 -1910.

Lewis (1839) — Lewis, Sir George Cornewall, *A Glossary of Provincial Words used in Herefordshire and some of the Adjoining Counties*, John Murray, London, 1839.

Lewis (1845) — Lewis, Samuel, *A Topographical Dictionary of England*, 4 Vols., 5th Edn., S. Lewis and Co., London, 1845.

LHEB — Jackson, Kenneth, *Language and History in Early Britain*, Edinburgh University Press, 1953.

Lhuyd — Lhuyd, E., *Archaeologia Britannica*, Oxford, 1707.

Liddall — Liddall, W. J. N., *The Place-Names of Fife and Kinross*, William Green & Sons, Edinburgh, 1896.

Lin. N&Q — *Lincolnshire Notes & Queries*, Vol. 2, Morton, Horncastle, 1891.

Lindsay (1921a) — Lindsay, W. M., (Ed.), *The Corpus, Epinal, Erfurt and Leyden Glossaries*, Oxford University Press, Oxford, 1921.

Lindsay (1921b)   Lindsay, W. M., (Ed.), *The Corpus Glossary*, with
                    an Anglo-Saxon index by Helen McM.
                    Buckhurst, The University Press,
                    Cambridge, 1921.

Lisle   Lisle, Edward, *Observations in Husbandry*, J.
                    Hughes (Printer), London, 1757.

LL   Evans, J. G. & Rhys, John, *The Text of the Book
                    of LLan Dav reproduced from the Gwysaney
                    Manuscript*, Oxford, 1893.

Lockwood   Lockwood, W. B., *A Panorama of Indo-European
                    Languages*, Hutchinson, London, 1972.

Long   Long, W. H., *A Dictionary of the Isle of Wight
                    Dialect*, Reeves & Turner, London and G. A.
                    Brannon & Co., Newport, 1886.

Longfellow   Longfellow, Henry Wadsworth, *Longfellow's
                    Poetical Works*, George Routledge and Sons,
                    London and New York, 1883.

Loth   Loth, Joseph, *Chrestomathie Bretonne*, Emile
                    Bouillon, Paris, 1890.

Loudon   Loudon, J. C., *An Encyclopædia of Agriculture*,
                    2nd Edn., Longman, Rees, Orme, Brown &
                    Green, London, 1831.

Lously   Lously, Major B., *A Glossary of Berkshire Words
                    and Phrases*, EDS, Trübner & Co., London,
                    1888.

M'Alpine   M'Alpine, Neil, *A Pronouncing Gaelic Dictionary*,
                    5th Edn., Maclachlan & Stewart, Edinburgh
                    and Simpkin, Marshall & Co., London,
                    1866.

MacBain (1911)   MacBain, Alexander, *An Etymological Dictionary
                    of the Gaelic Language*, Eneas MacKay,
                    Stirling, 1911.

MacBain (1922)   MacBain, Alexander, *Place-Names Highlands &
                    Islands of Scotland*, Eneas MacKay, Stirling,
                    1922.

Mackay   Mackay, John & Hereford, J.P., (Eds.)
                    Sutherland Place-Names, p 189, in
                    Transactions of the Gaelic Society of
                    Inverness, Vol. 18, Inverness, 1891-2.

Mackenzie   Mackenzie, Kenneth, *Lewis Place-Names*, pp

368-87, in Transactions of the Gaelic
Society of Inverness, Vol. 26, Inverness,
1910.

Mackintosh — Mackintosh, W. R. *Around the Orkney Peat-Fires*,
3rd Edn., Printed by W. R. Mackintosh,
Kirkwall, 1914.

MacLean — MacLean, L., *The History of the Celtic Language*,
Smith, Elder and Co., Edinburgh, 1840.

Macleod & Dewar — Macleod, Norman & Dewar, Daniel, *Dictionary
of The Gaelic Language*, John Grant,
Edinburgh, 1909.

Macray — Macray, W. D., *Notes from The Muniments of St.
Mary Magdelan College, Oxford.* Parker & Co.,
Oxford and London, 1882.

Madden — Madden, Sir Frederic, *Sir Gawain and the Green
Knight*, Richard and John E. Taylor (printer),
London, 1839.

Marryat — Marryat, Captain, *Olla Podrida*, George
Routledge, London, 1840.

Marshall (1787) — Marshall, William, *The Rural Economy of Norfolk*,
Vol. 2, printed for T Cadell, London, 1787.

Marshall (1796a) — Marshall, William, *The Rural Economy of the
Midland Counties*, Vol. 2, printed for G.
Nicol; G. G. and J. Robinson; and J.
Debrett, London, 1796.

Marshall (1796b) — Marshall, William, *East Yorkshire Words, 1796*,
in Reprinted Glossaries, Series B, Skeat,
The Rev. W. W., EDS, (Ed.), Trübner &
Co. Limited, London, 1879.

Marshall (1873a) — Marshall, W. H., *Provincialisms Of East Norfolk*,
in Reprinted Glossaries, by Skeat, The Rev.
W. W., (Ed.), EDS, Trübner & Co. Limited,
London, 1873.

Marshall (1873b) — Marshall, W.H., *Provincialisms Of West Devonshire*,
in Reprinted Glossaries, Skeat, The Rev. W.
W., (Ed.), EDS, Trübner & Co. Limited,
London, 1873.

Martin — Martin, Charles Trice, (Compiler)*The Record
Interpreter: A Collection of Abbreviations, Latin
Words and Names used in English Historical*

Manuscripts and Records, 2nd Edn., Stevens and Sons, London, 1910.

Marwick   Marwick, Hugh, *The Orkney Norn*, Oxford University Press, 1929 and Brinnoven, Livingston, 1995.

Massey   Massey, Gerald, *A Book of The Beginnings*, Williams & Norgate, London, 1881-3, and eres., available from www.masseiana.org/bkkk5.htm (accessed January 30th, 2015).

Matasovic   Matasovic, Ranko, *An Etymological Lexicon of Proto-Celtic*, eres, available from www.scribd.com (accessed 20th November 2014).

McClure   McClure, E., *British Place-Names in Their Historical Setting*, Society for the promotion of Christian Knowledge, London, 1910.

McKay   McKay, Patrick, *A Dictionary of Ulster Place-Names*, 2nd Edn., Cló Ollscoil na Banriona, Belfast, 2007.

MED   *Middle English Dictionary*, (Lewis, Robert E., Editor-in-Chief), University of Michigan, eres., available from http://quod.lib.umich.edu/m/med/ accessed to date).

Meyer   Meyer, Kuno, *Contributions To Irish Lexicography*, Vol. 1, Part 1, Max Niemeyer, Halle and David Nutt, London, 1906.

Middendorff   Middendorff, Heinrich, *Altenglisches Flurnamenbuch*, Max Niemeyer, Halle, 1902.

Mills   Mills, A. D., *A Dictionary of British Place-Names*, Oxford University Press, 2003.

Milne (1906)   Milne, John, *achnach* p 127, in *Scottish Notes and Queries*, , 2nd Series Vol. 7, The Rosemount Press, Aberdeen, 1906.

Milne (1912)   Milne, John, *Celtic Place-Names in Aberdeenshire*, Aberdeen Daily Journal, 1912.

Milne (undated) Milne, John, *Gaelic Place-Names of The Lothians*, M'dougall's Educational Company, Limited. Edinburg and London, undated.

MIPN   O'Laughlin, M. C., *The Master Book of Irish Placenames*, Irish Genealogical Foundation, Kansas, USA, 1994.

Moisy   Moisy, Henri, *Glossaire Comparatif Anglo-Normand*, Henri Delesques, Caen, 1889.

Moor   Moor, Edward, *Suffolk Words and Phrases, or an attempt to collect The Lingual Localisms of that County*, printed by J. Loder for R. Hunter, London, 1823.

Moore   Moore, A. W., *The Surnames & Place-Names of the Isle of Man*, Elliot Stock, London, 1890.

Morgan   Morgan, T., *The Place-Names of Wales*, 2nd Edn., John E. Southall, Newport, 1912.

Morris (1857)   Morris, Richard, *The Etymology of Local Names*, Judd & Glass, London, 1857.

Morris (1865)   Morris, The Rev. Richard, (Ed.), *The Story of Genesis and Exodus*, EETS, Trübner & Co. Limited, London 1865.

Morris (1869)   Morris, The Rev. Richard, (Ed.), *Early English Alliterative Poems in the West-Midland Dialect of the Fourteenth Century*, 2nd Edn., EETS, Trübner & Co. Limited, London, 1869.

Morris (1874)   Morris, The Rev. Richard, (Ed.), *Cursor Mundi (The Cursur o the World)*, 6 Vols., EETS, Kegan Paul, Trench, Trübner & Co. Limited, London, 1874-92.

Mortimer   Mortimer, J., *The Whole Art of Husbandry, or The Ways of Managing and Improving of Land*, Vol. 2, printed for R. Robinson and G. Mortlock, London, 1716.

Morton   Morton, J. C., *A Cyclopedia of Agriculture*, 2 Vols., Blackie & Son, London, 1856.

Moville   *Townland of Stroove*, PN derivation available from http: //www.movilleinishowen. com/travel/townlands/Stroove.htm (accessed 8th October 2013).

MR   *Museum Rusticum et Commerciale*, Vol. 2, Printed for R Davis, London, 1764.

MSP   Manx Society Publications.

MWDB   Market Weighton Drainage Board, *Adopted*

Watercourses, available from http://www.marketweighton-idb.org/adopted-watercourses.htm (accessed 8th November 2010).

Nance (1978)   Nance, R. Morton, *An English-Cornish and Cornish-English Dictionary*, Cornish Language Board, 1978.

Nance (1990)   Nance, R. Morton, *A New Cornish Dictionary*, Dyllansow Truran, Redruth, 1990.

Napier   Napier, Arthur S., *Old English Glosses Chiefly Unpublished*, The Clarendon Press, Oxford, 1900.

Nennius   Todd, James Henthorn, (Ed.), *The Irish Version of the Historia Britonum of Nennius*, The Royal Irish Archaeological Society, Dublin, 1848.

Newbury   The *History and Antiquities of Newbury and its Environs*, Hall & Marsh, London, 1839.

Nicolaisen   Nicolaisen, W .F .H., *Scottish Place-Names: Their Study and Significance*, Original Edn. Batsford, London, 1976, New Edn. J Donald (Birlinn Ltd), Edinburgh, 2001.

Nodal & Milner Nodal, John H. & Milner, George, *A Glossary of the Lancashire Dialect*, Manchester Literary Club Publication, Trübner & Co. Ltd and Alexander Ireland & Co., London, 1875.

Norris   Norris, Edwin, *Ancient Cornish Vocabulary* (pp 311-432) in *The Ancient Cornish Drama*, Vol. 2, OUP, Oxford, 1859.

Northall   Northall, G. F., *A Warwickshire Word-Book*, EDS, Henry Frowde, London, 1896.

Norwich (1892) Hudson, The Rev. William, (Ed.), *Leet Jurisdiction in the City of Norwich*, Selden Society Publications, Bernard Quaritch, London, 1892.

Norwich (1906) Hudson, The Rev. William & Tingey, John Cottingham (Eds.),*The Records of the City of Norwich*, Vol. 1, Jarrold & Sons, London and Norwich, 1906.

Norwich (1910) Tingey, John Cottingham, (Ed.), *The Records of the City of Norwich*, Vol. 2, Jarrold & Sons,

London and Norwich, 1910.

Nostratic — Dolgopolsky, Aharon, *Nostratic Dictionary*, McDonald Institute for Archaeological Research, Cambridge, 2008, eres., available from http://www.dspace.cam.ac.uk/handle /1810/196512 (accessed 9th May 2009).

O'Donovan (1856) — O'Donovan, John, (Ed.), *Annals of The Kingdom of Ireland, by The Four Masters*, Vol. 1, 2nd Edn., Hodges, Smith &Co., Dublin, 1856.

O'Donovan (1860) — O'Donovan, John, (Ed.), *Annals of Ireland. Three fragments*, The Irish Archæological and Celtic Society, University Press, Dublin, 1860.

O'Reilly — O'Reilly, Edward, *An Irish-English Dictionary*, with a supplement containing many thousand Irish words by John O'Donovan, James Duffy, Dublin & London 1864.

O'Brien — O'Brien, John, *Focaloir Gaoidhilge-Sax-Bhearla or An Irish-English Dictionary*, 2nd Edn. Hodges & Smith, Dublin, 1832.

OED1 — *The Oxford English Dictionary*, 1st Edn., 10 Vols., Originally published in fascicules based on *A New English Dictionary on Historical Principles; Founded Mainly on the Materials Collected by The Philological Society* and only later as *The Oxford Emglish Dictionary*, The The Clarendon Press, Oxford, 1888-1928 and eres., available from www.ned0.org (accessed to date).

OED2 — *The Oxford English Dictionary*, 2nd Edn., 20 Vols., The Clarendon Press, Oxford, 1989.

OED3 — *The Oxford English Dictionary*, 3rd Edn., 2010-, online eres., available by subscription, or public library card log in, from http://www.oed.com/loginpage

Ogilvie — Ogilvie, John, *The Imperial Dictionary of The English Language*, 4 Vols., Annandale, Charles, (Ed.) Blackie & Son, London, 1882-3.

OGS — *Ordnance Gazetteer of Scotland*, 6 Vols., Groome,

	Francis H., (Ed.), Thomas C. Jack, Edinburgh, 1884-5.
OIM	Old Irish Maps available from http://maps.osi.ie/publicviewer (accessed to date).
OLD	*The Oxford Latin Dictionary*, Fascicules (i-vi), The Clarendon Press, Oxford, 1968-80.
Oldfield	Oldfield, Edmund, *A Topographical and Historical Account of Wainfleet and the Wapentake of Candleshoe in the County of Lincoln*, Longman & Others, London, 1829.
Oliver	Oliver, the Younger, Stephen, *Rambles in Northumberland and on the Scottish Border*, Chapman & Hall, London, 1835.
OMLD	*Dictionary of Medieval Latin from British Sources*, Fascicules (i-viii), The Clarendon Press, Oxford, 1975-2003.
Oppenheimer	Oppenheimer, Stephen, *The Origins of the British*, Robinson, London, 2007.
OPr.EV	Palmaitis, Dr L., & Holcwesscher, P., *Old Prussian Elbing Vocabulary*, eres., available from http://donelaitis.vdu.lt/prussian/Elbin.pdf (accessed 25th November 2011).
Ormulum	Holt, The Rev. Robert (Ed.), *The Ormulum with the notes and glossary of Dr. R. M. White*, 2 Vols., The Clarendon Press, Oxford, 1878.
OSIDS	Ordnance Survey Ireland *Discovery* Series maps.
OSL	Ordnance Survey *Landranger* Series maps.
OSM	*Open Street Map*, search facilities available from http://www.openstreetmap.org (accessed to date).
OSMS	*Old Scottish Maps* available from http://www.scottish-places. info/ counties/ countymaps24.html (accessed 3rd October 2012).
OSNIDS	Ordnance Survey Ireland *Discoverer* Series maps.
OSOL	Ordnance Survey *Outdoor Leisure* Series maps.
OSX	Ordnance Survey *Explorer* Series maps.
Padel	Padel, O. J., *A Popular Dictionary of Cornish Place-*

Names, Alison Hodge, Penzance, 1988.

Page   Page, John Lloyd Warden, *An Exploration of Exmoor and the Hill Country of West Somerset*, Seeley and Co., Ltd, London, 1890.

Palmaitis   Palmaitis, Dr L., (Translator), *Trilingual Dictionary of Prussian* (Prussian, German, English) based upon *The Etymological Dictionary of Prussian* by Vytautas Mažiulis (Mažiulis, V., Prūsų kalbos etimologijos žodynas. Mokslas, Vilnius, I 1988, II 1993, III 1996, IV 1997) eres., available from http://poshka.bizland.com/prussian/recon structions.htm (accessed 25th November 2011).

Palsgrave (1852)   Palsgrave, Jean, *L'eclaircissement de La Langue Française*, Imprimerie Nationale, Paris, 1852.

Palsgrave (1896)   Palsgrave, The Rev. F. M. T., *A List of Words and Phrases in every-day use by the Natives of Hetton-Le-Hole in the County of Durham*, EDS, Henry Frowde, London, 1896.

Parish   Parish, The Rev. W. D., *A Dictionary of the Sussex Dialect and Collection of Provincialisms in use in the County of Sussex*, Farncombe & Co., Lewes, 1875.

Parish & Shaw   Parish, The Rev. W. D., & Shaw, The Rev. W. F., *A Dictionary of the Kentish Dialect and Provincialisms in use in the County of Kent*, Farncombe & Co., Lewes, 1888.

Parkinson   Parkinson, R., *General View of the Agriculture of the County of Huntingdon*, Board of Agriculture, printed for Richard Phillips, London, 1811.

Pashka   Joseph Pashka, J., *English-Sudovian Sudovian-English Dictionary*, eres., available from http://www.suduva.com/virdainas/ (accessed 25th November 2011).

Patterson   Patterson, William Hugh, *A Glossary of Words in use in the Counties of Antrim and Down*, EDS, Trübner & Co., London 1880

PC   *Proto-Celtic–English Wordlist* and *English–Proto-*

Celtic Word List together with an *English-Old Irish Annotated Word List* is available from http://www.wales.ac.uk/ by pasting Celtic-Lexicon into the search box (accessed 2008 to date).

Peacock (1869) Peacock, Robert Backhouse & Atkinson, The Rev. J. C., Ed., *A Glossary of the Dialect of the Hundred of Lonsdale*, in Transactions of The Philological Society, printed by Asher & Co., London, 1869.

Peacock (1870) Peacock, Edward, *Ralf Skirlaugh, The Lincolnshire Squire*, 3 Vols., Chapman & Hall, London, 1870.

Peacock (1889) Peacock, Edward, *A Glossary of Words used in the Wapentakes of Manley and Corringham, Lincolnshire*, 2nd Edn., 2 Vols., Trübner & Co. Limited, London, 1889.

Pegge Pegge, The Rev. S., *Two Collections of Derbicisms*, EDS, Henry Frowde, London, 1895-6.

Phillips (1720) Phillips, Edward, *The New World of Words or Universal English Dictionary*, 7th Edn., Kersey, John, (Ed.), London, 1720.

Phillips (1803) Phillips, J., *A General History of Inland Navigation*, 4th Edn., Printed for J. Taylor & C. & R. Baldwin, London, 1803.

Phillips (1853) Phillips, John, *The Rivers, Mountains and Sea-Coast of Yorkshire*, John Murray, London, 1853.

PNBd See PNBdHu.

PNBdHu Mawer, Allen & Stenton, F. M., *The Place-names of Bedfordshire & Huntingdonshire*, EPNS 3, Cambridge, 1926.

PNBk Mawer, Allen & Stenton, F. M., *The Place-Names of* Buckinghamshire, EPNS 2, Cambridge, 1925.

PNBrk Gelling, Margaret, *The Place-Names of Berkshire*, 3 Vols., EPNS 49-51, Cambridge, 1973-6.

PNC Reaney, P. H., *The Place-Names of Cambridgeshire and the Isle of Ely*, EPNS 19, Cambridge, 1943.

PNCh   Dodgson, John McN. & Rumble, A. R., *The Place-Names of Cheshire*, 7 Vols., EPNS 44-8, 54, (Dodgson, John McN.,) 74 (Dodgson, John McN. & Rumble, A. R.), Cambridge & Nottingham, 1970-98.

PNCu   Armstrong, A. M., Mawer, A., Stenton, F. M. & Bruce Dickins. *The Place-Names of Cumberland*, 3 Vols., EPNS 20-2, Cambridge, 1950-52.

PND   Gover J. E. B., Mawer A. & Stenton F. M., *The Place-Names of Devon*, 2 Vols., EPNS 8-9, Cambridge, 1931-2.

PNDb   Cameron, Kenneth, *The Place-Names of Derbyshire*, 3 Vols., EPNS 27-9, 1959

PNDo   Mills, A. D., *The Place-Names of Dorset*, 6 Vols., EPNS 52-3, 59-60 & 86-7, Nottingham, 1977-2010.

PNDu   Watts, Victor, *The Place-Names of County Durham*, EPNS 83, Nottingham, 2007.

PNEss   Reaney, P. H., *The Place-Names of Essex*, EPNS 12, Cambridge, 1935.

PNGBI   Field, J., *Place-Names of Great Britain and Ireland*, David & Charles, Newton Abbot and London, 1990.

PNGl   Smith, A. H., *The Place-Names of Gloucestershire*, 4 Vols., EPNS 38-41, Cambridge, 1964-5.

PNHrt   Gover J. E. B; Mawer A. & Stenton F. M., *The Place-Names of Hertfordshire*, EPNS 15, Cambridge, 1938.

PNHunts   *The Place-Names of Huntingdonshire*, see PNBdHu.

PNL   Cameron, Kenneth; Field, John; Insley, John & Cameron, Jean, *The Place-Names of Lincolnshire*, 8 Vols., EPNS 58, 64, 65 (Cameron, Kenneth) 66, 71,73, 77 (Cameron, Kenneth; Field, John; Insley, John) & 85 (Cameron, Kenneth; Insley, John & Cameron, Jean), Nottingham, 1985-2010.

PNLa   Ekwall, E., *The Place-Names of Lancashire*, Chetham Society Publications, New Series,

Vol. 81, Manchester, 1922.

PNLei   Cox, Barrie, *The Place-Names of Leicestershire*, 5
         Vols., EPNS 75, 78, 81, 84, 88, Nottingham,
         1998-2011.

PNMx   Gover J. E. B., Mawer A. & Stenton F. M.,
         with the collaboration of Madge, S. J., *The
         Place-Names of Middlesex, apart from the City of
         London*, EPNS 18, Cambridge, 1942.

PNNf   Sandred, Karl Inge., Lindström, Bengt.,
         Cornford, B., Lindström, B., Rutledge P. &
         Schram, O. K., *The Place-Names of Norfolk*, 3
         Vols., EPNS 61, (Sandred, Karl Inge;
         Lindström, Bengt) 72 (Sandred, Karl Inge.,
         Cornford, B., Lindström, B., Rutledge P. &
         Schram, O. K.) & 79 (Sandred, Karl Inge),
         Nottingham, 1989-2002.

PNNorDu Mawer A., *The Place-Names of Northumberland and
         Durham*, Cambridge University Press, 1920.

PNNt   Gover J. E. B., Mawer A. & Stenton F. M., *The
         Place-Names of Nottinghamshire*, EPNS 17,
         Cambridge, 1940.

PNNth   Gover J. E. B., Mawer A. & Stenton F. M., *The
         Place-Names of Northamptonshire*, EPNS 10,
         Cambridge, 1933.

PNO   Gelling, Margaret, *The Place-Names of
         Oxfordshire*, (based on material collected by
         Doris Mary Stenton), 2 Vols., EPNS 23-4,
         Cambridge, 1953-4.

PNRB   Rivet, A. L. F. & Smith, Colin, *The Place-Names
         of Roman Britain*, Batsford, London, 1979
         and Book Club Associates, London, 1981.

PNRu   Cox, Barrie, *The Place-Names of Rutland*, 3 Vols.
         in 1, EPNS 67-9, Nottingham, 1994.

PNSa   Gelling, Margaret, in collaboration with Foxall,
         H. D. G., *The Place-Names of Shropshire*, 7
         Vols., EPNS 62-3, 70, 76, 80, 82, 89,
         Nottingham, 1990-2012.

PNSr   Gover J. E. B., Mawer A. & Stenton F. M., in
         collaboration with A. Bonner, *The Place-
         Names of Surrey*, EPNS 11, Cambridge, 1934.

PNSt   Oakden, J. P., *The Place-Names of Staffordshire*, EPNS 55, Nottingham, 1984.

PNSx   Gover J. E. B., Mawer A. & Stenton F. M., *The Place-Names of Sussex*, 2 Vols., EPNS 6-7, Cambridge, 1929-30.

PNW   Gover J. E. B., Mawer A. & Stenton F. M., *The Place-Names of Wiltshire*, EPNS 16, Cambridge, 1939.

PNWa   Gover J. E. B., Mawer A. & Stenton F. M., in collaboration with F. T. S. Houghton, *The Place-Names of Warwickshire*, EPNS 13, Cambridge, 1936.

PNWe   Smith, A. H., *The Place-Names of Westmorland*, 2 Vols., EPNS 42-3, Cambridge, 1967

PNWo   Gover J. E. B., Mawer A. & Stenton F. M., in collaboration with F. T. S. Houghton, *The Place-Names of Worcestershire*, EPNS 4, Cambridge, 1927.

PNYE   Smith, A. H., *The Place-Names of the East Riding of Yorkshire and York*, EPNS 14, Cambridge, 1937.

PNYN   Smith, A. H., *The Place-Names of the North Riding of Yorkshire*, EPNS 5, Cambridge, 1928.

PNYW   Smith, A. H., *The Place-Names of the West Riding of Yorkshire*, 8 Vols., EPNS 30-37, Cambridge, 1961-63.

Pokorny   Pokorny, Julius, Indogermanisches Etymologisches Wörterbuch, A. Franke, Bern, 1959.

Polychronicon   *Polychronicon Ranulphi Higden Monachi Cestrensis ; together with the English Translations of John Trevisa and of an unknown writer of the fifteenth century*, Vol. 1, Babington, Churchill, (Ed.), Rolls Series, Longman, Green, Longman, Roberts, and Green, London, 1865.

PP   *Promptoriun Parvulorum, The First English-Latin Dictionary*, 1440 A.D., Mayhew, A. L., (Ed.), EETS, Extra Series No. 102, Kegan Paul, Trench, Trübner & Co. Limited and Henry Frowde, London, 1908.

Prevost   Prevost, E. W., *A Supplement to the Glossary of the*
                   Dialect of Cumberland, Henry Frowde;
                   Bemrose & Sons Ltd, London and C
                   Thurnam & Sons, Carlisle, 1905.
Price   Price, Glanville, *The Languages of Britain*,
                   Edward Arnold, London, 1985.
Prince   Prince, John Dyneley, *Materials for a Sumerian*
                   Lexicon, J. C. Hinrich'sche Buchhandlung,
                   Liepzig, 1908.
Pryce   Pryce, W., *Archaeologia Cornu-Britannica*,
                   Cruttwell (Printer), Sherborne, 1790.
Pughe & Pryse   Pughe, W. Owen & Pryse, Robert John (Ed.),
                   Geiriadur Cenhedlaethol Cymraeg a Saesneg, A
                   National Dictionary of the Welsh Language with
                   English and Welsh Equivalents, 3rd Edn., 2
                   Vols., Robert Gee, Denbigh, 1866 & 1873.
Pughe   Pughe, W. Owen, *A Dictionary of the Welsh*
                   Language, 2nd Edn, 2 Vols., Thomas Gee,
                   Denbigh, 1832.
Pulman (1857)   Pulman, George P. R., *Local Nomenclature. A*
                   Lecture on The Names of Places, Longman,
                   Brown, Green, Longmans & Roberts,
                   London, 1857
Pulman (1871)   Pulman, George P. R., *Rustic Sketches Being*
                   Rhymes and "Skits" on Angling and Other
                   Subjects, 3rd Edn., John Russel Smith,
                   London, 1871.
Pulman (1875)   Pulman, George P. R., *The Book of the Axe*, 4th
                   Edn., Longman, Green, Reader & Dyer,
                   London, 1875.
Ravenna   Parthey, G. & Pinder, M., (Eds.), *Ravennatis*
                   Anonymi Cosmographia Et Gvidonis Geographica,
                   Berolini, Berlin, 1860.
Ray   Ray, J., *North Country Words*, EDS, Series B,
                   XV-XVII, 1874.
Rees   Rees, Abraham, *The Cyclopædia or Universal*
                   Dictionary of Arts, Sciences, and Literature, 39
                   Vols., Longman, Hurst, Rees, Orme and
                   Brown, London, 1819.
Reid   Reid, Robert, *Old Glasgow and its Environs*,

	Historical and topographical, David Robertson, Glasgow and longman & Co., London, 1864.
Rempel	Rempel, H., *A Mennonite Low German Dictionary*, eres., available from http://www. mennolink.org/doc/lg/index.html (accessed 25th November 2011).
Rennie	Rennie, James, (Ed.), *The Magazine of Botany & Gardening*, Vol. 2, G Henderson, London, 1834.
RGSS	*The Register of the Great Seal of Scotland*, Vol. 6, Thomson, John Maitland, (Ed.) H. M. General Register House, Edinburgh, 1890.
Richards	Richards, W., *A Pocket Dictionary, Welsh-English*, R Hughes & Son, Wrexham and Simpkin, Marshall, and Co., London, 1861.
Riley (1859)	Riley, Henry Thomas, (Ed.) *Liber Albus*, in Vol. 1 of Munimeta Gildhallæ Londoniensis; Liber Albus, Liber Cusumarum et Liber Horn, Longman, Brown, Green, Longmans and Roberts, London, 1859.
Riley (1860)	Riley, Henry Thomas, (Ed.) *Liber Cusumarum*, in Vol. 2, Parts 1 & 2, of Munimeta Gildhallæ Londoniensis; Liber Albus, Liber Cusumarum et Liber Horn, Longman, Green, Longman, And Roberts, London, 1860.
Riley (1862)	Riley, Henry Thomas, (Ed.) *Translation of the Anglo-Norman passages in Liber Albus, Glossaries, Appendices, and Index*, in Vol. 3 of Munimeta Gildhallæ Londoniensis; Liber Albus, Liber Cusumarum et Liber Horn, Longman, Green, Longman, and Roberts, London, 1862.
Ritson	Ritson, Joseph, *Ancient Engleish Metrical Romanceës*, Vol. 2, printed by W. Bulmer & Co., for G & W Nicol, Booksellers, London, 1802.
Robertson	Robertson, J. Drummond, *A Glossary of Dialect and Archaic Words used in the County of*

Gloucester, Lord Moreton (Ed.) EDS, Kegan Paul, Trench, Trübner & Co., London 1896.

Robinson (1855)   Robinson, F. K., *A Glossary of Yorkshire Words and Phrases, collected in Whitby and the Neighbourhood*, John Russell Smith, London, 1855.

Robinson (1876a)   Robinson, F. K., *A Glossary of Words used in the Neighbourhood of Whitby*, Trübner & Co., London 1876.

Robinson (1876b)   Robinson, C. Clough, *A Glossary of Words pertaining to the Dialect of Mid-Yorkshire with others peculiar to Lower Nidderdale*, EDS, Trübner & Co., London, 1876.

Room   Room, A., *A Dictionary of Irish Place-Names*, The Appletree Press Ltd, Belfast, 1986.

Ross (1877)   Ross, Frederick; Stead, Richard & Holderness, Thomas, *A Glossary of Words used in Holderness in the East Riding of Yorkshire*, EDS, Trübner & Co., London, 1877.

Ross (2001)   Ross, D., *Scottish Place-Names*, Birlinn Ltd., Edinburgh, 2001.

Rotzoll   Rotzoll, Eva, *Die Deminutivbildungen im Neuenglischen unter besonderer Berücksichtigung der Dialekte*, Carl Winter's Universitätsbuchhandlung, Heidelberg, 1910.

Rowe   Rowe, Samuel, *A Perambulation of the Antient and Royal Forest of Dartmoor and the Venville Precincts*, 3rd Edn., revised and corrected by J. Brooking Rowe. James G. Commin, Exeter and Gibbings & Co., Ltd, London, 1896.

Ruskin   Ruskin, John, *Præterita*, Vol 1., George Allen, London, 1907.

Rye   Rye, Walter, *A Glossary of Words used in East Anglia. Founded on that of Forby*, EDS, Henry Frowde, London, 1895

S   Sawyer, P. H., *Anglo-Saxon Charters: an Annotated List and Bibliography*, Royal Historical Society, London, 1968 and eres., commonly known as *The Electronic Sawyer*, available from http://www.esawyer.org.uk /about/index.html (accessed 2008 to date).

Salesbury   Salesbury, Wyllyam, *A Dictionary in Englyshe and Welshe*, original published by John Waley, 1547, reprinted for the Cymmrodorion Society by T. Richards, London, 1877.

Salisbury   Salisbury, Jesse, *A Glossary of Words and Phrases used in S.E. Worcestershire*, J. Salisbury, London, 1893.

Sandys   Sandys, George, *The Poetical Works of George Sandys*, now first collected with introductions and notes by The Rev. Richard Hooper, 2 Vols., John Russell Smith, London, 1872.

Saywell   Saywell, The Rev. J. L., *The History and Annals of Northallerton, Yorkshire*, Simpkin, Marshall & Co., London, 1885.

SED   *Survey of English Dialects - The Dictionary and Grammar*, Upton, C., Parry, D., & Widdowson, J.D.A. (Eds,), Routledge, London, 1994.

Seward   Seward, Wm. Wenman, *Topographia Hibernica*, (unpaginated) Alex. Stewart (Printer), Dublin, 1797.

Sibbald   Sibbald, Sir Robert, *Description of the Islands of Orkney and Zetland*, reprinted from the edition of 1711 by Robert Monteith, Thomas G. Stevenson, Edinburgh, 1845.

Sinclair   Sinclair, Sir John, *An Account of the Systems of Husbandry*, 2nd Edn., Vol. 1, James Ballantyne & Co., (Printer), Edinburgh, 1813.

Skeat (1886)   Skeat, The Rev. W. W., (Ed.), *The Wars of Alexander: An Alliterative Romance*, EETS, Extra Series, Vol. 47, Trübner & Co., London, 1886.

Skeat (1888)   Skeat, The Rev. W. W., *An Etymological Dictionary of the English Language*, The Clarendon Press, Oxford, 1888.

Skeat (1896)	Skeat, The Rev. W. W., *Nine Specimens of English Dialects*, Henry Frowde, London, 1896.
Skene (1867)	Skene, William F., (Ed.), *Chronicles of the Picts, Chronicles of the Scots, and other early memorials of Scottish History*, H. M. General Register House, Edinburgh, 1867.
Skene (1868)	Skene, William F., (Ed.), *The Four Ancient Books of Wales*, 2 Vols., Edmonston & Douglas, Edinburgh, 1868.
Skinner	Skinner, Stephen, *Etymologicon Linguæ Anglicanæ*, London, 1671
SLB (1905)	*Court Leet Records, 1550-1577*, Vol. 1, Pt. 1, Hearnshaw, F. J. C. & D. M. (Eds.), Southampton Record Society, H. M. Gilbert & Son, Southampton, 1905.
SLB (1906)	*Court Leet Records, 1578-1602*, Vol. 1, Pt. 2, Hearnshaw, F. J. C. & D. M. (Eds.), Southampton Record Society, H. M. Gilbert & Son, Southampton, 1906.
SLB (1907)	*Court Leet Records, 1603-1624*, Vol. 1, Pt. 3, Hearnshaw, F. J. C. & D. M. (Eds.), Southampton Record Society, Cox & Sharland, Southampton, 1907.
SLB (1908)	*Glossary*, pp 3-51 in *Supplement to Court Leet Records, 1550-1624*, Masom, W. F. (Ed.), Southampton Record Society, Cox & Sharland, Southampton, 1908.
Slow	Slow, Edward, *The Wiltshire Moonraker's edition of West Countrie Rhymes*, R. R. Edwards, Salisbury & Simpkin, Marshall; Hamilton, Kent & Co. Ltd, London, 1903.
Smart & Crofton	Smart, M. D. & Crofton, H. T., *The Dialect of the English Gypsies*, 2nd Edn., Asher & Co., London, 1875.
Smiles (1861-2)	Smiles, Samuel, *Lives of The Engineers*, 3Vols., Vols 1 & 2 1861, Vol 3, 1862, John Murray, London, 1861.
Smiles (1864)	Smiles, Samuel, *James Brindley and The Early Engineers*, Abridged from *Lives of The Engineers*, John Murray, London, 1864.

Smiles (1878)	Smiles, Samuel, *George Moore Merchant and Philanthropist*, 2nd Edn., George Routledge & Sons, London, 1878.
Smith (1851)	Smith, Robert, *Some Account of the Formation of Hillside Catch-meadows on Exmoor*, pp 139-148, in The Journal of The Royal Agricultural Society of England, Vol. 4, John Murray, London, 1851 and pp 448-453 in The British Farmer's Magazine, Vol. 20, Henry Wright, London 1851.
Smith (1881)	Smith, Major Henry & Smith, C. Roach, *Isle of Wight Words*, in Original Glossaries, Series C, EDS, The Rev. W. W. Skeat, (Ed.), Trübner & Co., London, 1881.
SND	*The Scottish National Dictionary*, 10 Vols., The Scottish National Dictionary Association Ltd, Edinburgh, 1931-1976 and eres., available from http://www.dsl.ac.uk/ (accessed 2005 to date).
Speight	Speight, Harry, *Romantic Richmondshire*, Elliot Stock, London 1897.
Spenser	Spenser, Edmund, The Faerie Queene, George Routledge, London, 1843.
Spurdens	Spurdens, The Rev. W. T., *East-Anglian Words, 1840*, in Reprinted Glossaries, Series B, Skeat, The Rev. W. W., (Ed.), EDS, Trübner & Co. Limited, London, 1879.
Spurrell's	*Geiriadur Cenhedlaethol Cymraeg a Saesneg, Spurrell's Welsh-English Dictionary*, 12th Edn., Anwyl, J. Bodvan (Ed.) W. Spurrell & Son, Camarthen, 1934.
SSR	Wrottesley, Major-General the Hon. George, (Ed.), *The Exchequer Subsidy Roll of A. D. 1327*, in Collections for a History of Staffordshire, Vol. 7, The William Salt Archæological Society, Harrison and Sons, London, 1886.
Stephens (1848)	Stephens, Henry, *A Manual of Practical Draining*, 3rd Edn., William Blackwood & Sons, Edinburgh and London, 1848.

Stephens (1889) Stephens, Henry, *The Book of the Farm*, in six divisions, 4th Edn., revised, and In great part rewritten, by James Macdonald. William Blackwood & Sons, Edinburgh and London, 1889-91.

Sternburg Sternburg, Thomas, *The Dialect and Folk-lore of Northamptonshire*, John Russell Smith, London, Abel & Sons; G. N. Wetton, Northampton, R. Todd, Oundle and A. Green, Brackley, 1851.

Stokes (1887) Stokes, Whitley, (Ed.), *The Tripartite Life of Patrick*, HMSO, London, 1887.

Stokes (1901 & 3) Stokes, Whitley & Strachan, John (Eds.), *Thesaurus Palaeohibernicus*, 2 Vols, Cambridge University Press, Vol. 1, 1901 & Vol. 2, 1903.

Stone Stone, Thomas *A review of the Corrected Agricultural Survey of Lincolnshire by Arthur Young*, 2nd Edn., Board of Agriculture, London, 1800.

Stratmann Stratmann, Francis Henry, *A Middle-English Dictionary*, new edition, revised & enlarged by Henry Bradley. The Clarendon Press, Oxford, 1891.

Sweet Sweet, Henry, (Ed.), *The Oldest English Texts*, EETS, N. Trübner & Co., London, 1885.

Sylvester Grosart, The Rev. Alexander B., (Ed.), *The Complete Works of Joshua Sylvester*, 2 Vols., Edinburgh University Press, 1880.

Taylor Taylor, Isaac, *Words and Places*, 2nd Edn, Macmillan and Co., London and Cambridge, 1865.

Thomas Thomas, Joseph, *Randigal Rhymes and a Glossary Of Cornish Words*, F. Rodda, Penzance, 1895.

Thompson Thompson, Pishey, *The History and Antiquities of Boston*, John Noble Jun., Boston; Logman & Co and Simpkin & Co., London and Samuel G. Drake, Boston, Massachusetts, 1856.

Thorpe (1835) Thorpe, Benjamin, (Ed.), *Libri Psalmorum Versio*

Antiqua Latina cum Paraphrasi Anglo-Saxonica, E Typographeo Academico, Oxford, 1835.

Thorpe (1861)   Thorpe, Benjamin, (Ed.), *The Anglo-Saxon Chronicle According to the Several Original Authorities*, 2 Vols., Longman, Green, Longman, Roberts, London, 1861.

Tipping   Tipping, Henry Avray, *In English Homes*, Vol. 2., George Newnes, London, 1908.

Turner (1880)   Turner, William H., *Selections from the Records of the City of Oxford*, James Parker & Co., Oxford and London, 1880.

Turner (1901)   Turner, J. Horsfall, *Yorkshire Place-Names, as recorded in the Yorkshire Domesday Book 1086*, T Harrison & Sons (Printer), Bingley, 1901.

Turner (1962)   Turner, R. L., *A Comparative Dictionary of the Indo-Aryan Languages*, Oxford University Press, London, 1962-6.

Turner (1969)   Turner, Dorothy Rivers, Indices to (Turner, R. L., *A Comparative Dictionary of the Indo-Aryan Languages*, Oxford University Press, London, 1962-6), Oxford University Press, London, 1969.

USBG (Bel.)   United States of America *Gazetteer of Belgium*, Board on Geographic Names, Dept. of the Interior, Washington 25, DC, 1963 and eres., available from http://geonames.nga.mil/ggmagaz/ (accessed 2010 to date). See USBG GNS source.

USBG (Den.)   United States of America *Gazetteer of Denmark*, Board on Geographic Names, Dept. of the Interior, Washington 25, DC, 1961 and eres., available from http://geonames.nga.mil/ggmagaz/ (accessed 2010 to date). See USBG GNS source.

USBG (Fra.)   eres., available from http://geonames.nga.mil/ggmagaz/ (accessed 2010 to date)

USBG (GDR)   United States of America *Gazetteer of the German Democratic Republic*, Board on Geographic

	Names, Dept. of the Interior, Washington 25, DC, 1983 and eres., available from http://geonames.nga.mil/ggmagaz/ (accessed 2010 to date). See USBG GNS source.
USBG (GFR)	United States of America *Gazetteer of the German Federal Republic*, Board on Geographic Names, Dept. of the Interior, Washington 25, DC, 1960 and eres., available from http://geonames.nga.mil/ggmagaz/ (accessed 2010 to date). See USBG GNS source.
USBG (GNS)	There are no licensing requirements or restrictions in place for the use of the GNS data. However, we recommend using the following citation to identify the GNS as a source: Toponymic information is based on the Geographic Names Database, containing official standard names approved by the United States Board on Geographic Names and maintained by the National Geospatial-Intelligence Agency. More information is available at the Products and Services link at www.nga.mil. The National Geospatial-Intelligence Agency name, initials, and seal are protected by 10 United States Code Section 425. See all USBG entries.
USBG (Ice.)	United States of America *Gazetteer of Iceland*, Board on Geographic Names, Dept. of the Interior, Washington 25, DC, 1961 and eres., available from http://geonames.nga.mil/ggmagaz/ (accessed 2010 to date). See USBG GNS source.
USBG (Lux.)	eres., available from http://geonames.nga.mil/ggmagaz/ (accessed 2010 to date). See USBG GNS source.
USBG (Nth.)	United States of America *Gazetteer of the Netherlands*, Board on Geographic Names, Dept. of the Interior, Washington 25, DC,

1950 and eres., available from http://
geonames.nga.mil/ggmagaz/ (accessed
2010 to date). See USBG GNS source.

USBG (Nwy.) eres., available from http://geonames.nga.
mil/ggmagaz/ (accessed 2010 to date). See
USBG GNS source.

USBG (Swe.) United States of America *Gazetteer of Sweden*,
Board on Geographic Names, Dept. of the
Interior, Washington 25, DC, 1989 and
eres., available from http: // geonames.
nga.mil/ggmagaz/ (accessed 2010 to date).
See USBG GNS source.

Vancouver (1795) Vancouver, Charles, *General View of the
Agriculture of the County of Essex*, Board of
Agriculture, London, 1795.

Vancouver (1808) Vancouver, Charles, *General View of the
Agriculture of the County of Devon*, Board of
Agriculture, London, 1808.

VCH H&IOW *The Victoria History of Hampshire and the Isle of
Wight*, Vol. 3, Page, William (Ed.),
Archibald Constable and Co. Ltd., London,
1908.

Vennemann Vennemann, Thomas, *Water all over the place:
The Old European toponyms and their Vasconic
origin*, available from
http://vennemann.userweb.mwn.de/#abst
racts (accessed 3rd September, 2004).

VI Visit Ireland glossary, available from
http://www.visitireland.com/planning/glo
ssary.asp (accessed 20th May, 2011).

Waddell Waddell, L. A., *A Sumer-Aryan Dictionary*, Part
1, A–F, Luzac & Co, London, 1927.

Wallenburg Wallenburg, Johannes Knut, *The Place-Names of
Kent*, Uppsala, 1934.

Walters Walters, The Rev. John, *An English and Welsh
Dictionary*, 3rd Edn., 2 Vols., Clwydian Press,
Denbigh, 1828.

Wardell Wardell, James, *The Municipal History of the
Borough of Leeds*, Longman, Brown & Co.,
London, 1846.

Warrack   Warrack, Alexander, *A Scots Dialect Dictionary*,
                        W & R Chambers Limited, Edinburgh &
                        London, 1911.

Watson (1904)   Watson, W. J., *Place-Names of Ross and Cromarty*,
                        The Northern Counties Printing and
                        Publishing Company, Limited, Inverness;
                        Norman Macleod, Edinburgh and David
                        Nutt, London, 1904.

Watson (1908)   Watson, W.J., *Topographical Varia*, p 148 in The
                        Celtic Review, Vol. 5, July 1908 to April
                        1909, Norman Macleod, Edinburgh, David
                        Nutt, London, & Hodges, Figgis & Co. Ltd,
                        Dublin.

Watson (1911)   Watson, W.J., *Topographical Varia V* , pp 361-
                        370 in The Celtic Review, Vol. 7, February
                        1911 to January 1912, William Hodge &
                        Co., Edinburgh, David Nutt, London, &
                        Hodges, Figgis & Co. Ltd, Dublin.

Watts   Watts, Victor, *The Cambridge Dictionary of English
                        Place-Names*, Cambridge University Press,
                        2010.

Weatherhill   Weatherhill, Craig, *A Concise Dictionary of
                        Cornish Place-Names*, Evertype, Westport,
                        2009.

Wells   Wells, Samuel, *The History of the Drainage of the
                        Great Level of the Fens, called Bedford Level*, 2
                        Vols., R. Pheney, London, 1830.

Wey   Wey, William, *The Itineraries of William Wey*, J. B.
                        Nichols and Sons, London, 1857

Wikipedia   *Mill-race* and *Water-meadow*, eres., available from
                        en.wikipedia.org (accessed 25th August,
                        2014).

Wiktionary   *Wiktionary, the free dictionary* available from
                        en.wiktionary.org (accessed continually to
                        date).

Wilbraham   Wilbraham, Roger, *An Attempt at a Glossary of
                        some Words used in Cheshire*, 2nd Edn,
                        reprinted from Archæologia Vol. XIX,
                        Richard Taylor (Printer), London, 1826.

Williams (1865)   Williams, R., *Lexicon Cornu-Britannicum*, Roderic,

Llandovery and Trübner & Co., London, 1865.

Williams (1872) Williams, Monier, *A Sanscrit-English Dictionary*, The Clarendon Press, Oxford, 1872.

Williams (1990) Williams, J E Caerwyn, *Wysg (river-name)*, wysg, hwysgynt, rhwysg, pp 670-8 in Celtica – Journal of the School of Celtic Studies - Vol. 21, 1990, available from http:// www.dias.ie/images/stories/celtics/pubs/celtica/c21/c21-670.pdf (accessed 18th May, 2010).

Williamson Williamson, W. A., *Local Etymology*, James Steel, Carlisle and Longman, Brown, Green and Longmans, London, 1849.

Willis Willis, Robert & Clarke, John Willis, (Ed.), *The Architectural History of The University Of Cambridge, and of The Colleges of Cambridge and Eton*, 4 Vols., The University Press, Cambridge, 1886.

Wilson (1855) Wilson, John, *Noctes Ambrosianæ*, Vol.1, William Blackwood & Sons, Edinburgh & London, 1855.

Wilson (1882) Wilson, The Rev. John, *The Gazeteer of Scotland*, W & A. K. Johnston, Edinburgh, 1882.

Windisch Windisch, Ernst, *Irische Texte mit Wörterbuch*, S. Hirzel, Leipzig, 1880.

Wordsworth Wordsworth, The Rev. Christopher, *Rutland Words*, EDS, Kegan Paul, Trübner & Co., London 1891.

Wright (1790) Wright, The Rev. T., *The Advantages and Method of Watering Meadows by Art*, 2nd Edn., S. Ruder (Printer), Cirencester, 1790.

Wright (1880) Wright, Thomas, *Dictionary of Obsolete and Provincial English*, 2 Vols., George Bell & Sons, London, 1880-1886.

WW Wright, Thomas & Wulcker, Richard Paul (Ed.), *Anglo-Saxon and Old English Vocabularies*, 2nd Edn., 2 Vols., Trübner & Co., London 1884.

WWS *The World's Writing Systems*, Daniels, P. T., & Blight, W. (Eds.) Oxford University Press,

1996.

Wycliffe   *The Holy Bible, containing the Old and New*
                  Testaments, with the Apocryphal Books, in the
                  Earliest English Versions made from the Latin
                  Vulgate by John Wycliffe and his followers. 4 vols.,
                  Forshall, The Rev. Josiah, and Madden, Sir
                  Frederic, (Eds.), Oxford University Press,
                  1850.

Wyntoun   Wyntoun, Androw of, *The Orygynale Cronykil of*
                  Scotland, Laing, David, (Ed.),Vols. 1 & 2,
                  Edmonston and Douglas, Edinburgh, 1872;
                  Vol. 3, William Paterson, Edinburgh, 1879.

YGM   *Y Geiriadur Mawr, The Complete Welsh-English &*
                  English-Welsh Dictionary, Evans, H.M., &
                  Thomas, W.O., Davis, Llandysul, 1978.

Young (1797)   Young, Arthur, *General View of the Agriculture of*
                  the County of Suffolk, Board of Agriculture,

Young (1799)   Young, Arthur, *General View of the Agriculture of*
                  the County of Lincoln, Board of Agriculture,
                  London,1799.

Zeuss   Zeuss, J. C., *Grammatica Celtica*, 2 Vols.,
                  Weidmann, Leipzig, 1853.

Appendix

Avon River (Upper Case) and Generic (Lower case) Forms inclusive of dual terms

An	afon	Auinus	awn
ab	Afonæ	aun	Awne
aba	Afone	Aune	Awni
Abae	Afvam	auney	awon
abainn	amain	auon	awy
abann	aman	Auona	Awyne
Abbona	amhain	auonn	Eafen
Abbone	amhainn	Auren	Evan
Aben	amhan	Ausona	Haefe
abh	amhann	Autona	Hafene
abha	amhuinn	Auvona	Hauene
abhainn	amon	Auwon	Havene
abhan	amuin	avan	Inn
abhann	an	Avele	Inney
abhin	Anne	Aven	Ive
abhoinn	aon	Avena	ob
abhuinn	au	Avene	oba
Abne	auan	Avenne	obadh
Abon	aub	Avern	obann
abona	Aue	avin	obha
Abonam	Auele	Avon	obhuin
abone	Auena	Avonam	oin
Abonis	Auenæ	Avone	othain
Æfene	auenam	Avonia	othainn
Aeron	Auene	Avren	oub
Afen	Auenes	Avyne	Oure
Afena	Auenina	Avynne	ow
Afenam	Auenna	aw	owen
Afenan	Auennus	awan	owin
Afene	Aueres	awin	owinn
Afne	Aufona	Awmon	Owne

www.ingramcontent.com/pod-product-compliance
Lightning Source LLC
Chambersburg PA
CBHW072301200526
45168CB00014B/98